International Conference on Modern Perspectives of Cosmology and Gravitation

(COSGRAV 12)

Journal of Physics: Conference Series Volume 405

Kolkata, West Bengal, India
7-11 February 2012

ISBN: 978-1-62276-871-4
ISSN: 1742-6588

Printed from e-media with permission by:

Curran Associates, Inc.
57 Morehouse Lane
Red Hook, NY 12571

Printed by Curran Associates, Inc. (2013)

For permission requests, please contact the Institute of Physics at the address below.

Institute of Physics
Dirac House, Temple Back
Bristol BS1 6BE UK

Phone: 44 1 17 929 7481
Fax: 44 1 17 920 0979

techtracking@iop.org

Additional copies of this publication are available from:

Curran Associates, Inc.
57 Morehouse Lane
Red Hook, NY 12571 USA
Phone: 845-758-0400
Fax: 845-758-2634
Email: curran@proceedings.com
Web: www.proceedings.com

TABLE OF CONTENTS

PREFACE

This volume of *Journal of Physics: Conference Series* is dedicated to the International Conference on Modern Perspectives of Cosmology and Gravitation (COSGRAV 12) organized by Physics and Applied Mathematics Unit, Indian Statistical Institute, Kolkata, held on 7–11 February 2012. The conference, which focused exclusively on recent trends of research in Cosmology and Gravitation, was the first in the series held in this institute of great repute. The ultimate plan is to make it a regular event every two or three years based on the very positive response we received which was beyond our expectation.

The immediate purpose of this conference was to bring together experienced as well as young scientists who are interested in working actively on various aspects of Cosmology and Gravitation. The lectures addressed major theoretical issues, current and forthcoming observational data as well as upcoming ideas in both theoretical and observational sectors. Keeping in mind the 'academic exchange first' approach the lectures were arranged in such a way that the young researchers had ample scope to interact with the stalwarts who are internationally leading experts in their respective fields of research.

The major topics covered in the conference are:

- Early Universe: Inflation, Alternatives and Links to Fundamental Physics
- Present Universe: Dark Matter, Dark Energy, Alternatives
- Observational Cosmology: CMB, Supernovae, Lensing, Galaxies and Clusters
- Quantum Aspects of Gravity
- Black Hole Physics
- Interface of Gravitation with Information Theory and Condensed Matter Physics.

Besides the invited talks a good proportion of the participants also presented their work through contributory talks and posters on this big platform. This was particularly encouraging and of benefit to the young participants, given that there were a number of scientists of international repute among the participants, the feedback from whom could guide them in the right direction. All the contributions were refereed by experts. This set a standard of its own.

We are indebted to the Indian Statistical Institute for providing us with generous funding that covered all the expenses required to organize such a huge conference, and for providing us with the support staff facilities. We gratefully acknowledge encouragement from Professor Bimal K Roy, Director, Indian Statistical Institute, and his constant support in all aspects of the conference which made the program function so well. We thank the Scientific Advisory Committee for their valuable suggestions on technical aspects of the conference. We thank all the members of Local Organizing Committee as well as the volunteering students who contributed their hard labour to make the conference a great success. Special thanks to Sudipta Das, Barun Kumar Pal and Sayantan Choudhury for their help during every stage of the conference.

We sincerely thank IOP Publishing and the staff of *Journal of Physics: Conference Series* for the publication of this issue.

Last but not least, we thank all the speakers and participants without whom the program would not have been such as success. We hope we will your active participation in future versions of the conference as well.

Supratik Pal and Banasri Basu (Editors)
Physics and Applied Mathematics Unit
Indian Statistical Institute
203 B.T.Road
Kolkata 700108
INDIA

Peer review statement

All papers published in this volume of *Journal of Physics: Conference Series* have been peer reviewed through processes administered by the proceedings Editors. Reviews were conducted by expert referees to the professional and scientific standards expected of a proceedings journal published by IOP Publishing.

Journal of Physics: Conference Series **405** (2012) 012001 doi:10.1088/1742-6596/405/1/012001

Quantum gravity, space-time structure, and cosmology

Martin Bojowald
The Pennsylvania State University, University Park, PA, USA
E-mail: bojowald@gravity.psu.edu

Abstract. A set of diverse but mutually consistent results obtained in different settings has spawned a new view of loop quantum gravity and its physical implications, based on the interplay of operator calculations and effective theory: Quantum corrections modify, but do not destroy, space-time and the notion of covariance. Potentially observable effects much more promising than those of higher-curvature effective actions result; loop quantum gravity has turned into a falsifiable framework, with interesting ingredients for new cosmic world views. At Planckian densities, space-time disappears and is replaced by 4-dimensional space without evolution.

Can we expect quantum-gravity observations? Even during inflation — the most energetic cosmological regime to which we have at least some indirect access — the Planck length $\ell_{\rm P}$ is far removed, by many orders of magnitude, from the cosmological scale $\ell_{\mathcal{H}} = 1/\mathcal{H}$, the inverse Hubble parameter. Any higher-curvature action

$$S_{\rm higher\ curvature} = \frac{1}{16\pi G}\int {\rm d}^4x\sqrt{-\det g}(\Lambda + R + A\ell_{\rm P}^2R^2 + \cdots)\,, \qquad (1)$$

often expected as the universal low-energy manifestation of quantum gravity, is then all but identical to the Einstein–Hilbert action. No sizeable effects of higher-order terms in the expansion should appear. Quantum gravity might be significant before inflation, at even higher energy densities, but then most effects are likely washed away during inflation and afterwards.

There is a more optimistic alternative: quantum gravity might modify space-time structure. An action covariant under the deformed symmetries of modified space-time would not be of the standard higher-curvature form. New possibilities for quantum-gravity effects arise. General features of quantized gravity show that such a scenario is to be expected: Gravitational dynamics is equivalent to space-time structure; evolving forward in time means probing different parts of space-time, from different perspectives. Quantum corrections in Hamiltonians, governing evolution of space-time and matter, could then imply modifications of space-time structure.

Space-time structure can be encoded algebraically and geometrically. An observer moving at constant speed v follows the trajectory $x(t) = x_0 + vt$. According to special relativity, we assign new coordinates

$$x' = \frac{x - vt}{\sqrt{1 - v^2/c^2}} \quad , \quad ct' = \frac{ct - vx/c}{\sqrt{1 - v^2/c^2}}$$

to events as seen by the moving observer. A spatial slice t = constant is transformed to a slanted slice t' = constant obeying the equation $ct = (v/c)x$ + constant in the old coordinates; Fig. 1.

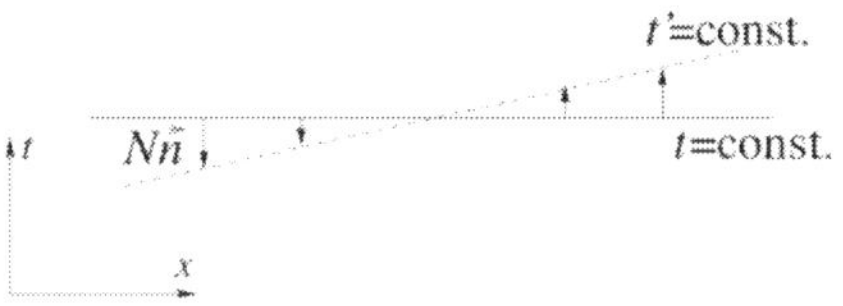

Figure 1. A Lorentz boost deforms the spatial slice t = constant in a linear fashion, the slope of the new slice determined by v/c.

Just as boosts, all Poincaré transformations can be realized geometrically as linear deformations of spatial slices, with functions $N(\vec{x}) = c\Delta t + (\vec{v}/c)\cdot\vec{x}$ implementing time translations and boosts by normal deformations, and spatial vector fields $\vec{w}(\vec{x}) = \Delta\vec{x} + \mathbf{R}\vec{\mathbf{x}}$ for spatial translations and rotations within a spatial slice. Also their commutators, that is the whole Poincaré algebra, can be geometrized. As an example, we look at the commutator of boosts and time translations: In Fig. 2, we see a normal deformation by $N_1(x) = vx/c$ (a boost with velocity v) followed by a normal deformation by $N_2(x) = c\Delta t - vx/c$ (the reverse Lorentz boost, and waiting Δt), performed in the two possible orderings. As elementary geometry shows, a boost of velocity v commutes with a time translation by Δt up to a spatial displacement $w(x) = \Delta x = v\Delta t$.

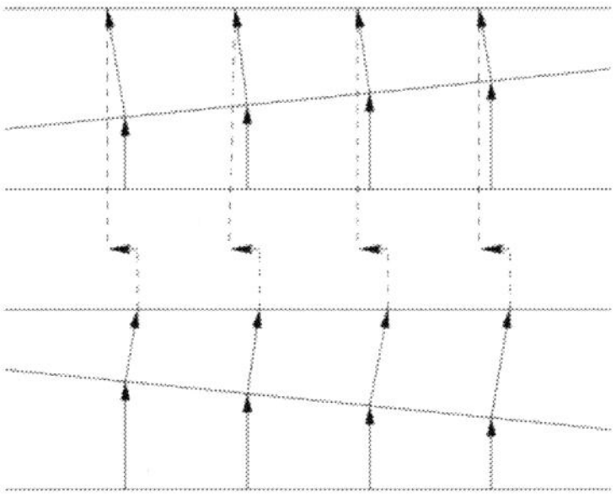

Figure 2. Commutator of a boost and a time translation. In the top part, we first perform a boost by v, then wait some time Δt and undo the boost; in the bottom part, the first boost is by $-v$, undone after time Δt. The image points after both deformations lie on the same spatial slice in both orderings, but are displaced by $\Delta x = v\Delta t$: During the time interval Δt, all objects have moved with velocity v.

We have geometrized uniform, inertial motion. In order to extend these constructions to general relativity, we must implement local Lorentz transformations: linear deformations with locally changing slope. Our transformations become arbitrary, non-linear coordinate changes — geometrically, we consider non-linear deformations of space. There are now infinitely many generators, given by spatial deformations $S(\vec{w}(x))$ along (shift) vector fields $\vec{w}$, and normal timelike deformations $T(N(x))$ by (lapse) functions $N(x)$. Geometry in diagrams such as Fig. 3, generalizing Fig. 2, produces the hypersurface-deformation algebra

$$[S(\vec{w}_1), S(\vec{w}_2)] = S(\mathcal{L}_{\vec{w}_2}\vec{w}_1) \quad (2)$$

$$[T(N), S(\vec{w})] = T(\vec{w}\cdot\vec{\nabla}N) \quad (3)$$

$$[T(N_1), T(N_2)] = S(N_1\vec{\nabla}N_2 - N_2\vec{\nabla}N_1)\,. \quad (4)$$

The last equation, via gradients $\vec{\nabla}$, makes use of a spatial metric.

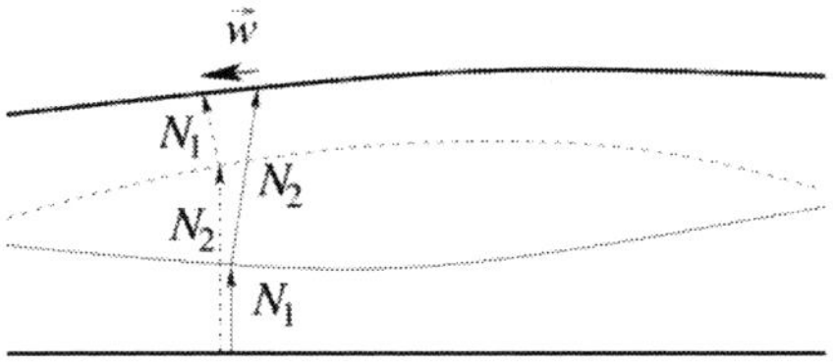

Figure 3. Non-linear hypersurface deformations. Repeated deformations along normals, in different orderings, differ by spatial shifts $\vec{w}$. Geometry implies the commutator (4). (See e.g. [1] for a detailed calculation.)

Hypersurface deformations geometrize the symmetries of space-time. Symmetry often determines dynamics. In the case of hypersurface deformations, a theory invariant under them is generally covariant [2]. As shown by [3, 4], second-order field equations for geometry (a spatial metric tensor), invariant under the hypersurface-deformation algebra as gauge transformations, must equal Einstein's. Classically, these are canonical expressions of the well-known fact that the Einstein–Hilbert action, as a 2-parameter family with Newton's and the cosmological constant, is unique among all second-order generally covariant actions (1) for the space-time metric. (For more details on canonical gravity, see [5].)

The geometrized view does not tell us much new about classical gravity. But it turns out to be useful for quantum gravity: One of the symmetries is a time translation, generated by the Hamiltonian. The Hamiltonian is of prime interest for quantized dynamics, and usually receives quantum corrections. If we quantize gravity and amend the classical Hamiltonian by quantum corrections, generators of the hypersurface-deformation algebra, and perhaps even the algebra itself, change. In the geometrized view, the question of quantum gravity, in very general terms, can be formulated thus: How does quantum physics change hypersurface deformations?

Loop quantum gravity is perhaps the best-developed canonical approach to quantum gravity. Not coincidentally, it has produced results about the hypersurface-deformation algebra. However, these features are often hidden because they require a proper treatment of the off-shell algebra (2)–(4), without fixing the space-time gauge or deparameterizing. Only recently have such structures become accessible.

The theory starts with a classical formulation of gravity more akin to Yang–Mills theories, describing space-time geometry by su(2)-valued "electric fields" $\vec{E}_i$ and "vector potentials" $\underset{\rightarrow}{A}_i$. The electric field is a densitized triad, determining spatial distances and angles by three orthonormal vectors $\vec{E}_i$, $i = 1, 2, 3$, at each point in space. The vector potential, the Ashtekar–Barbero connection, is a combination of different measures of curvature of space. (Arrows above the letter indicate contravariant vectors; arrows underneath, covariant ones. If the metric is one of the basic fields to be quantized, here written in terms of $\vec{E}_i$, the tensorial type of fields must be kept track of, for a metric factor inserted to pull up or down indices could significantly change operator properties after quantization.)

The great advantage of these fields, compared to the spatial metric and its rate of change or extrinsic curvature, is that they offer natural smearings, of $\underset{\rightarrow}{A}_i$ along curves (holonomies) and $\vec{E}_i$ over surfaces (fluxes). Some kinematical divergences — delta functions in Poisson brackets — can be eliminated, allowing one to represent both fields as operators. By analogy with the harmonic oscillator or Fock-space methods in quantum field theory, holonomies then play the role of creation operators by which one can construct the state space, and fluxes, some kind of number operator, determine the excitation level.

In what follows, we use a U(1)-connection $\underset{\rightarrow}{A}$ to illustrate salient features. Holonomies then read $h_e = \exp(i \int_e \mathrm{d}\lambda \underset{\rightarrow}{A} \cdot \vec{t}_e)$, integrated along curves e in space with tangent $\vec{t}_e$. Starting with a simple basic state ψ_0, defined by $\psi_0(\underset{\rightarrow}{A}) = 1$, we construct excited states

$$\psi_{e_1,k_1;\ldots;e_i,k_i}(\underset{\rightarrow}{A}) = \hat{h}_{e_1}^{k_1} \cdots \hat{h}_{e_i}^{k_i} \psi_0(\underset{\rightarrow}{A}) = \prod_e h_e(\underset{\rightarrow}{A})^{k_e} = \prod_e \exp(ik_e \textstyle\int_e \mathrm{d}\lambda \underset{\rightarrow}{A} \cdot \vec{t}_e) \,. \qquad (5)$$

Collecting all curves used in a graph $g = \{e_1, \ldots, e_i\}$, we view these states and their superpositions as discrete lattice-works, or discrete quantum space, illustrated in Fig. 4. The starting state ψ_0 has no excitations, not even of geometry. It is emptier than the matter vacuum, emptier than empty space.

Flux operators, constructed from $\vec{E}$ canonically conjugate to $\underset{\rightarrow}{A}$, are derivative operators

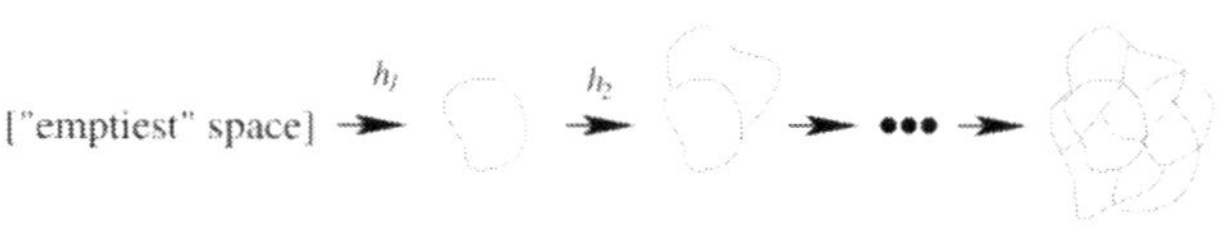

Figure 4. Holonomy operators generate discrete, lattice-like geometry out of "emptiest" space.

$\int_S \mathrm{d}^2y\underset{\to}{n}\cdot\hat{\vec{E}}$ integrated over surfaces S in space with co-normal $\underset{\to}{n}$. They acy by

$$\int_S \mathrm{d}^2y\underset{\to}{n}\cdot\hat{\vec{E}}\psi_{g,k} = -8\pi i G\hbar\int_S \mathrm{d}^2y\underset{\to}{n}\cdot\frac{\delta\psi_{g,k}}{\delta\underset{\to}{A}(y)} = 8\pi\ell_{\mathrm{P}}^2\sum_{e\in g}k_e\mathrm{Int}(S,e)\psi_{g,k} \tag{6}$$

with the intersection number $\mathrm{Int}(S,e)$ and the Planck length $\ell_{\mathrm{P}} = \sqrt{G\hbar}$. These operators have discrete spectra, implying discrete geometry [6, 7]: for gravity, fluxes represent the spatial metric. The elementary structure of space is realized at the Planck scale $\ell_{\mathrm{P}} \sim 10^{-35}\mathrm{m}$, at least kinematically, but may be excited to larger sizes $\sqrt{k}\ell_{\mathrm{P}}$ with integer k.

It remains challenging to formulate a consistent version of discrete dynamics, for instance of cosmic expansion, but some general features are known [8]. The classical Hamiltonian of gravity depends on $\underset{\to}{A}_i$, quantized by holonomies. Every one of these operators acts by creating new geometrical excitations along curves e, or new lattice sites in a discrete picture. The changing excitation level of geometry is measured by fluxes.

Yang–Mills theory on Minkowski space-time has a Hamiltonian $H_{\text{Yang-Mills}} = \kappa\int \mathrm{d}^3x(|\vec{E}_i|^2 + |\vec{B}_i|^2)$ for $\vec{B}_i = \underset{\to}{\nabla}\times\underset{\to}{A}_i + C_{ijk}\underset{\to}{A}_j\times\underset{\to}{A}_k$, with structure constants C_{ijk}. Gravity on any space-time makes use of similar fields, but in a more complicated, non-polynomial way:

$$H_{\text{gravity}} = \frac{1}{16\pi G}\int \mathrm{d}^3x\frac{\sum_{ijk}\epsilon_{ijk}(\vec{B}_i\times\vec{E}_j)\cdot\vec{E}_k}{\sqrt{\frac{1}{6}|\sum_{ijk}\epsilon_{ijk}(\vec{E}_i\times\vec{E}_j)\cdot\vec{E}_k|}} + \cdots \tag{7}$$

with $C_{ijk} = \epsilon_{ijk}$.

This Hamiltonian implies characteristic corrections when quantized in the loopy way. We have already noted that holonomies are used to quantize $\underset{\to}{A}_i$ (or $\vec{B}_i$), replacing linear or quadratic functions with non-polynomial ones. Expanding holonomies $h_e(\underset{\to}{A}_i)$ in Taylor series, one can match the leading terms with the classical expression, but higher-order terms remain as corrections motivated by quantum geometry. Flux operators are linear in $\vec{E}_i$, but they imply corrections nonetheless: Flux operators (6) have discrete spectra containing zero and therefore lack inverse operators, but classically we must divide by a function of $\vec{E}_i$ to compute the Hamiltonian (7). It turns out that Hamiltonian operators can be defined [8, 9], in a more round-about way circumventing direct inverses of fluxes. Additional quantum corrections result.

One can illustrate these features by a simpler system: inverse momentum on a circle with canonical variables (φ, p). We have states $\langle\varphi|n\rangle = \exp(in\varphi)$ for all integer n, a momentum operator $\hat{p}|n\rangle = \hbar n|n\rangle$, and "$\hat{p}^{-1}$" is not densely defined: it would be infinite on $|0\rangle$. Instead, we use the classical identity

$$I := \frac{1}{2}|p|^{-1/2}\mathrm{sgn}(p) = \cos(\varphi)\{\sin(\varphi),\sqrt{|p|}\} - \sin(\varphi)\{\cos(\varphi),\sqrt{|p|}\}$$

to quantize $\hat{I} = -i\hbar^{-1}\left(\widehat{\cos}\left[\widehat{\sin},\sqrt{|\hat{p}|}\right] - \widehat{\sin}\left[\widehat{\cos},\sqrt{|\hat{p}|}\right]\right)$ as a well-defined operator with the correct classical limit at large n, in regimes with $\hbar n = p_n$ finite for $\hbar\to 0$.

Applying this procedure to the gravitational Hamiltonian, we have inverse-triad corrections from quantizing

$$\left\{\underset{\to}{A}^i, \int\sqrt{|\det E|}\mathrm{d}^3x\right\} = 2\pi G\epsilon^{ijk}\frac{\vec{E}_j\times\vec{E}_k}{\sqrt{|\det E|}}\mathrm{sgn}(\det E)\,.$$

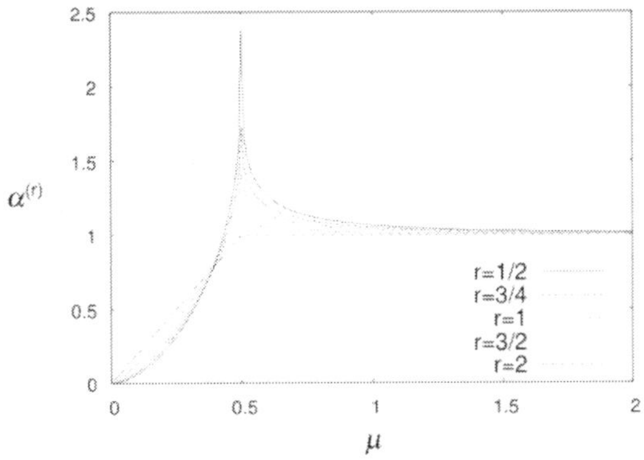

Figure 5. Inverse-triad corrections $\alpha = \langle \hat{I} \rangle(\mu)/I(\mu)$: the ratio of quantum expectation value and classical inverse in terms of flux eigenvalues μ. (The parameter r is an example for a quantization ambiguity.)

As illustrated in Fig. 5, using the calculations for quantizations and expectation values developed in [10, 11, 12], an automatic cut-off of $1/E$-divergences at degenerate triads results, but also corrections for large flux values. In addition, we have higher-order corrections from using holonomies for $\vec{B}_i$, and, as always in interacting quantum theories, quantum back-reaction of fluctuations and higher moments of a state on expectation values.

All these corrections modify the classical dynamics. Inverse-triad corrections, for instance, enter the Hamiltonian by

$$\frac{1}{16\pi G}\int \mathrm{d}^3x\, \alpha(\vec{E}_l)\, \frac{\sum_{ijk}\epsilon_{ijk}(\vec{B}_i\times\vec{E}_j)\cdot\vec{E}_k}{\sqrt{\frac{1}{6}|\sum_{ijk}\epsilon_{ijk}(\vec{E}_i\times\vec{E}_j)\cdot\vec{E}_k|}} + \cdots$$

The Hamiltonian, in turn, generates time translations as part of the hypersurface-deformation algebra. Computing algebraic relations for the modified Hamiltonian, ensuring that a closed algebra is still realized (the quantization is then anomaly-free), one finds that the hypersurface-deformation algebra does change [13]. It is modified, but not broken:

$$[S(\vec{w}_1), S(\vec{w}_2)] = S(\mathcal{L}_{\vec{w}_2}\vec{w}_1) \tag{8}$$

$$[T_\alpha(N), S(\vec{w})] = T_\alpha(\vec{w}\cdot\vec{\nabla}N) \tag{9}$$

$$[T_\alpha(N_1), T_\alpha(N_2)] = S(\alpha^2(N_1\vec{\nabla}N_2 - N_2\vec{\nabla}N_1))\,. \tag{10}$$

By inverse-triad corrections we deform but do not violate covariance. Consistent deformations of the same form have by now been derived in different models and with various methods, including effective constraints of spherical symmetry [14, 15, 16, 17] and operator calculations in $2+1$-dimensional and other models [18].

Some immediate consequences can be seen by specializing the algebra to linear N and $\vec{w}$, comparing with Fig. 2: The relation between spatial displacements and boost velocities is modified: $\Delta x = \alpha^2 v\Delta t$, with $\alpha \neq 1$ in general. For fluxes larger than the Planck area, $\alpha > 1$ according to Fig. 5: discrete space speeds up propagation, much like discrete matter structures can change propagation speeds of light or phonons. We are dealing with a new form of space-time structure, a form which we cannot yet handle directly in the absence of a manifold picture, perhaps a non-commutative one. Several subtleties remain in the usual concepts employed to analyze effects of general relativity, especially the notion of horizons [19]. But some consequences, for instance in cosmological perturbation equations, can be analyzed based only on the modified S and T_α: These functions tell us how fields transform under changes of perspective in space-time — we are able to derive gauge-invariant perturbations — and how they evolve when we move forward in time using T_α [20].

The modified dynamics of density perturbations u and gravitational waves w, the two gauge-invariant fields of most interest, can be condensed to [21]

$$-u'' + s(\alpha)^2\Delta u + (\tilde{z}''/\tilde{z})u = 0 \quad , \quad -w'' + \alpha^2\Delta w + (\tilde{a}''/\tilde{a})w = 0 \tag{11}$$

with functions $\tilde{a}$, $\tilde{z}$ and s all related to the correction function α derived from inverse-triad operators. Different modes propagate with different speeds, giving rise to corrections to the tensor-to-scalar ratio and other potentially observable effects. The theory can be tested, and it is falsifiable: $\delta := \alpha - 1$ is large for small lattice spacing (see Fig. 5), while large lattice spacings imply strong discretization effects and holonomy corrections. One can estimate that $\delta > 10^{-8}$ is required by these theoretical arguments of mutual consistency [21]. Observations give upper bounds on δ: We have two-sided bounds on the discreteness scale, not just an upper bound that could always be evaded by tuning parameters. An analysis [12, 22] using the currently available observational data provides the upper bound $\delta < 10^{-4}$, Fig. 6. While four orders of magnitude still remain to be bridged to rule out the theory, the bounds are much closer to each other than the Planck length is to the cosmic Hubble scale, an immense range in which pure curvature corrections would be viable. For additional observational results regarding tensor modes, see [23, 24, 25]

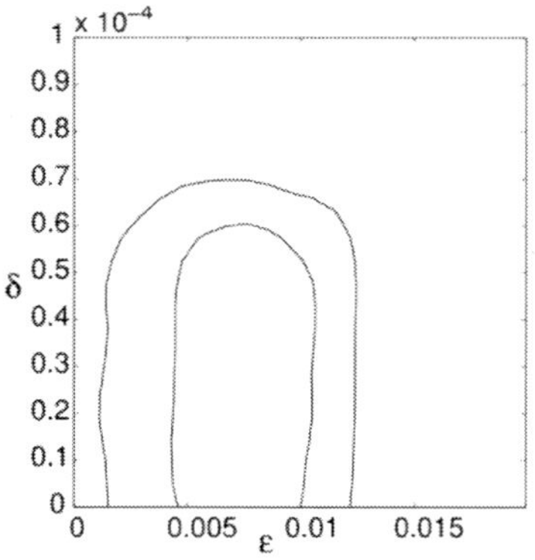

Figure 6. Likelihood plot for allowed values of the slow-roll parameter ϵ and the inverse-triad correction δ. Observations are consistent with $\delta = 0$ and provide an upper bound $\delta < 10^{-4}$.

Inverse-triad corrections, depending on the microscopic discreteness scale of a state in relation to the Planck length, are the main correction in sub-Planckian curvature regimes. Closer to the big bang, where the Planck density looms, all corrections, including holonomy effects, are important. Many surprising consequences have been explored in models of loop quantum cosmology [26, 27].

Discreteness means that the wave function of the universe is subject to a difference equation [28, 29],

$$C_+(\mu)\psi_{\mu+1} - C_0(\mu)\psi_\mu + C_-(\mu)\psi_{\mu-1} = \hat{H}_{\text{matter}}(\mu)\psi_\mu \tag{12}$$

with wave fuctions in the triad representation, depending on flux eigenvalues $\mu \in \mathbb{R}$. Following the recurrence, a state is extended uniquely through the classical singularity at $\mu = 0$. The singularity is resolved by this mechanism of quantum hyperbolicity [30]. In the deep quantum regime, holonomy corrections and dynamical quantum effects are strong and it is difficult to find an intuitive geometrical picture for the resolution mechanism. Fortunately, some properties can be seen in solvable, harmonic models [31]: the state bounces, reaching a non-zero minimal volume close to the Planck density [32]. This consequence is realized in a special model in which no quantum back-reaction occurs, a flat isotropic model with a free massless scalar. Approximately, the same behavior is realized for models with curvature or small matter masses and self-interactions, but the scenario remains to be confirmed under more general situations [33, 34].

Journal of Physics: Conference Series **405** (2012) 012001 doi:10.1088/1742-6596/405/1/012001

Even though the singularity is resolved, some features demonstrate that space-time, if it even exists in a meaningful form in those extremes, behaves in surprising ways. One of these features is cosmic forgetfulness [35, 36], addressing the question of how much pre-big bang information can be recovered after a state has moved through a resolved classical singularity. Fig. 7 shows sample solutions, the spread around the central lines indicating quantum fluctuations. Different fluctuations can be realized before and after the high-density phase, quantitatively expressed by a squeezing parameter of states [37].

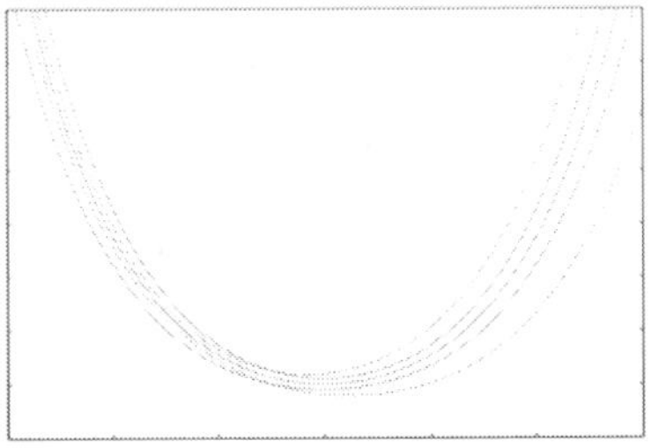

Figure 7. Dynamical wave functions illustrated by their expectation values (central lines) and fluctuations (spreads). The expectation value always bounces at some minimal volume V in this harmonic holonomy-modified model, but fluctuations on the two sides can differ in significant ways.

In more detail, the ratio of volume fluctuations ΔV at late and early times t is bounded by

$$\left|1 - \frac{\Delta V_{t\to-\infty}}{\Delta V_{t\to\infty}}\right| \lesssim \frac{\Delta H_{\rm matter}/\langle \hat{H}_{\rm matter}\rangle}{(\Delta V/\langle \hat{V}\rangle)_{t\to\infty}}, \tag{13}$$

derived with semiclassical techniques and dynamical coherent states [37, 36, 38]. The right-hand side of this inequality is typically large: the validity of quantum field theory on curved space-time at $t \to \infty$ indicates that matter-energy fluctuations are much larger than geometry fluctuations. The ratio $\Delta V_{t\to-\infty}/\Delta V_{t\to\infty}$, and evolutionary aspects influenced by it, remains largely unconstrained, even if the classical singularity can be resolved.

Space-time at and before the big bang is not necessarily classical, even if isotropic models remain well-defined. Additional information about space-time can be gained from the hypersurface-deformation algebra, now evaluated with holonomy corrections for strong-curvature regimes. Preliminary results in different models [14, 39, 40, 1] (taking into account partial holonomy effects) indicate that the algebra is again deformed:

$$[S(\vec{w}_1), S(\vec{w}_2)] = S(\mathcal{L}_{\vec{w}_2}\vec{w}_1) \tag{14}$$

$$[T_\beta(N), S(\vec{w})] = T_\beta(\vec{w}\cdot\vec{\nabla}N) \tag{15}$$

$$[T_\beta(N_1), T_\beta(N_2)] = S(\beta(N_1\vec{\nabla}N_2 - N_2\vec{\nabla}N_1)) \tag{16}$$

with a function β depending on $\underset{\rightarrow}{A}_i$, or on curvature. In contrast to inverse-triad corrections, holonomy corrections can result in $\beta < 0$, which they do at high density (the "bounce"). In constructions such as Fig. 2, a velocity in one direction would imply motion in the opposite direction, rather intransigent behavior. More meaningfully, one can interpret the phenomenon as signature change [1]: Euclidean 4-dimensional space has a hypersurface-deformation algebra with $\beta = -1$. What appears as a "bounce" in simple isotropic models is not a bounce at all; there is no evolution and no fully deterministic relation between pre- and post-big bang states.

Loop quantum gravity implies radical modifications at Planckian densities, with Euclidean space instead of space-time. There is no propagation of structure from collapse to expansion, as assumed in bounce models. Instead, loop quantum cosmology shows a non-singular beginning of the Lorentzian phase, when the diluting density falls sufficiently below the Planck density. Such a place, nonsingular and yet unprecedented, is natural for posing initial conditions, for instance for an inflaton state. We obtain a mixture of linear and cyclic models. On less extreme scales, loop quantum gravity implies space-time, but with modified structures, allowing quantum-gravity corrections more significant than higher-curvature terms. Inverse-triad corrections are not very far from testability; loop quantum gravity is falsifiable.

This work was supported in part by NSF grant PHY0748336.

References:

[1] Bojowald M and Paily G M (*Preprint* arXiv:1112.1899)
[2] Dirac P A M 1958 *Proc. Roy. Soc. A* **246** 333–343
[3] Hojman S A, Kuchař K and Teitelboim C 1976 *Ann. Phys. (New York)* **96** 88–135
[4] Kuchař K V 1974 *J. Math. Phys.* **15** 708–715
[5] Bojowald M 2010 *Canonical Gravity and Applications: Cosmology, Black Holes, and Quantum Gravity* (Cambridge: Cambridge University Press)
[6] Rovelli C and Smolin L 1995 *Nucl. Phys. B* **442** 593–619, erratum: *Nucl. Phys. B* **456** 753 (*Preprint* gr-qc/9411005)
[7] Ashtekar A and Lewandowski J 1997 *Class. Quantum Grav.* **14** A55–A82 (*Preprint* gr-qc/9602046)
[8] Thiemann T 1998 *Class. Quantum Grav.* **15** 839–873 (*Preprint* gr-qc/9606089)
[9] Thiemann T 1998 *Class. Quantum Grav.* **15** 1281–1314 (*Preprint* gr-qc/9705019)
[10] Bojowald M 2001 *Phys. Rev. D* **64** 084018 (*Preprint* gr-qc/0105067)
[11] Bojowald M, Hernández H, Kagan M and Skirzewski A 2007 *Phys. Rev. D* **75** 064022 (*Preprint* gr-qc/0611112)
[12] Bojowald M, Calcagni G and Tsujikawa S 2001 *JCAP* **11** 046 (*Preprint* arXiv:1107.1540)
[13] Bojowald M, Hossain G, Kagan M and Shankaranarayanan S 2008 *Phys. Rev. D* **78** 063547 (*Preprint* arXiv:0806.3929)
[14] Reyes J D 2009 *Spherically Symmetric Loop Quantum Gravity: Connections to 2-Dimensional Models and Applications to Gravitational Collapse* Ph.D. thesis The Pennsylvania State University
[15] Bojowald M, Reyes J D and Tibrewala R 2009 *Phys. Rev. D* **80** 084002 (*Preprint* arXiv:0906.4767)
[16] Kreienbuehl A, Husain V and Seahra S S 2012 *Class. Quantum Grav.* **29** 095008 (*Preprint* arXiv:1011.2381)
[17] Kreienbuehl A, Husain V and Seahra S S (*Preprint* arXiv:1109.3158)
[18] Henderson A, Laddha A and Tomlin C (*Preprint* arXiv:1204.0211)
[19] Bojowald M, Paily G M, Reyes J D and Tibrewala R 2011 *Class. Quantum Grav.* **28** 185006 (*Preprint* arXiv:1105.1340)
[20] Bojowald M, Hossain G, Kagan M and Shankaranarayanan S 2009 *Phys. Rev. D* **79** 043505 (*Preprint* arXiv:0811.1572)
[21] Bojowald M and Calcagni G 2011 *JCAP* **1103** 032 (*Preprint* arXiv:1011.2779)
[22] Bojowald M, Calcagni G and Tsujikawa S 2011 *Phys. Rev. Lett.* **107** 211302 (*Preprint* arXiv:1101.5391)
[23] Barrau A and Grain J 2009 *Phys. Rev. Lett.* **102** 081301 (*Preprint* arXiv:0902.0145)
[24] Grain J, Cailleteau T, Barrau A and Gorecki A 2010 *Phys. Rev. D* **81** 024040 (*Preprint* arXiv:0910.2892)
[25] Mielczarek J and Szydłowski M 2007 *Phys. Lett. B* **657** 20–26 (*Preprint* arXiv:0705.4449)
[26] Bojowald M 2008 *Living Rev. Relativity* **11** 4 http://www.livingreviews.org/lrr-2008-4 (*Preprint* gr-qc/0601085)
[27] Bojowald M 2011 *Quantum Cosmology: A Fundamental Theory of the Universe* (New York: Springer)
[28] Bojowald M 2001 *Phys. Rev. Lett.* **86** 5227–5230 (*Preprint* gr-qc/0102069)
[29] Bojowald M 2002 *Class. Quantum Grav.* **19** 2717–2741 (*Preprint* gr-qc/0202077)
[30] Bojowald M 2007 *AIP Conf. Proc.* vol 910 pp 294–333 proceedings of the XIIth Brazilian School on Cosmology and Gravitation (*Preprint* gr-qc/0702144)
[31] Bojowald M 2007 *Phys. Rev. D* **75** 081301(R) (*Preprint* gr-qc/0608100)
[32] Ashtekar A, Pawlowski T and Singh P 2006 *Phys. Rev. Lett.* **96** 141301 (*Preprint* gr-qc/0602086)
[33] Bojowald M 2008 *Gen. Rel. Grav.* **40** 2659–2683 (*Preprint* arXiv:0801.4001)
[34] Bojowald M 2008 *Phys. Rev. Lett.* **100** 221301 (*Preprint* arXiv:0805.1192)
[35] Bojowald M 2007 *Nature Physics* **3** 523–525
[36] Bojowald M 2008 *Proc. Roy. Soc. A* **464** 2135–2150 (*Preprint* arXiv:0710.4919)
[37] Bojowald M 2007 *Phys. Rev. D* **75** 123512 (*Preprint* gr-qc/0703144)
[38] Kaminski W and Pawlowski T 2010 *Phys. Rev. D* **81** 084027 (*Preprint* arXiv:1001.2663)
[39] Perez A and Pranzetti D 2010 *Class. Quantum Grav.* **27** 145009 (*Preprint* arXiv:1001.3292)
[40] Cailleteau T, Mielczarek J, Barrau A and Grain J (*Preprint* arXiv:1111.3535)

Journal of Physics: Conference Series **405** (2012) 012002 doi:10.1088/1742-6596/405/1/012002

Bulk–brane models: an overview and some queries

Sayan Kar

Department of Physics & Centre for Theoretical Studies, Indian Institute of Technology Kharagpur 721 302, India.

E-mail: sayan@iitkgp.ac.in

Abstract. In this article we first present a concise overview, highlighting the main features of the well-known bulk–brane models. We then discuss specific topics related to Lorentz violations, localisation of fields on the brane and braneworld gravity. Finally, we describe, in brief, a lower dimensional analog model for the warped two-brane system and conclude with some queries and remarks.

1. A very brief review

Almost a century ago, in 1914, Nordstrom [1] first introduced the notion of dimensions beyond the usual four. A few years later, pioneering work by Kaluza and Klein (KK) [2] laid the theoretical foundations of unification using compact extra dimensions. The KK idea, though pursued in terms of detail by many, remained largely dormant for quite some time till it reappeared and gained support mainly through the work of string theorists [3], during the second half of the twentieth century. In the late 1990s, another new, phenomenological idea arrived on the scene whose origins may be traced to early work by Akama, Rubakov–Shaposhnikov, Visser and Gogberashvilli [4]. This new class of extra dimensional models which are significantly different from the Kaluza–Klein models are known today, collectively, as the *bulk-brane* scenario [5, 6].

1.1. Jargon

The term *bulk*, in these models, refers to the background higher dimensional spacetime whereas *brane* (derived from mem*brane*) refers to an embedded $3+1$ dimensional, timelike hypersurface in the bulk. We assume we live on one such *brane*. The presence of the extra dimensions is manifest in phenomena on the brane. Advantages of the bulk-brane constructions include possibilities of (a) resolution of some well–known problems (b) verification of the existence of extra dimensions in collider and other terrestrial experiments (eg. non–Newtonian short distance forces) or astrophysical observations.

There are two types of bulk–brane models: the LED (large extra dimensional) model due to ADD [5] and the WED (warped extra dimensional) models due to RS [6]. We focus henceforth on the WED models, where we have two varieties: a *two-brane* model with an orbifolded extra dimension (topology S^1/Z_2, finite and noncompact) and a *single brane* model where the extra dimension is of infinite extent. Fig. 1 illustrates the above notions, qualitatively.

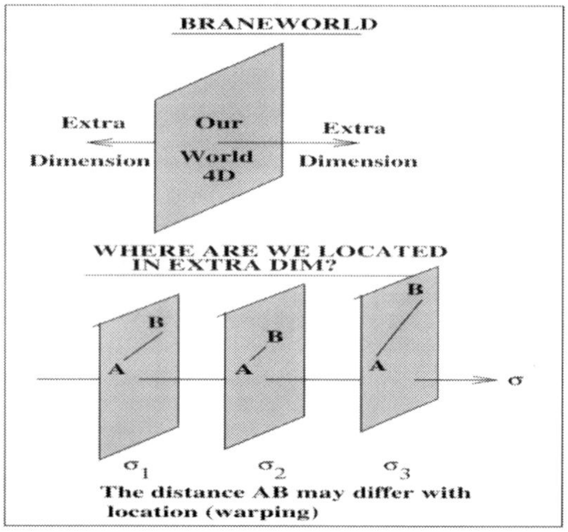

Figure 1. Bulk and brane, several branes, warping

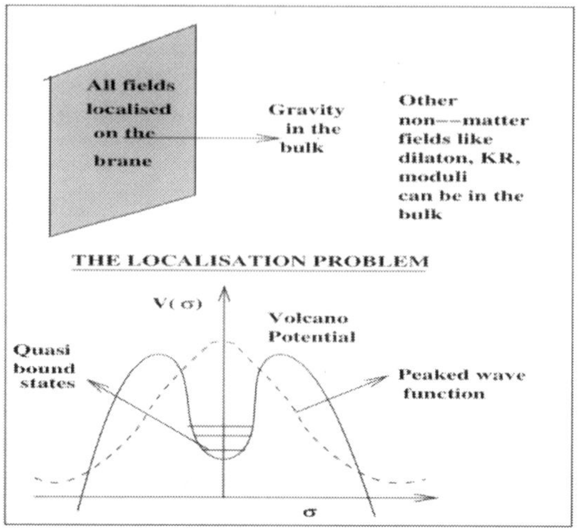

Figure 2. Localisation of fields on the brane, volcano potential.

1.2. *Hierarchy*

The hierarchy problem, in a loose sense, asks why the electroweak (100 GeV) and Planck (10^{19} GeV) scales are so widely separated? In the LED models, one does answer this, though a new hierarchy is generated [5]. The RS model resolution is better and does not introduce a new hierarchy. Briefly, one begins with the following five dimensional *warped* metric ansatz

$$ds^2 = e^{-2k|\sigma|}\left(-dt^2 + |d\vec{r}|^2\right) + d\sigma^2 \quad (1)$$

where σ denotes the extra dimension and $e^{-2k|\sigma|}$ is known as the warp factor. A σ =constant hypersurface is the *brane*. The bulk geometry here is AdS (constant negative scalar curvature), generated by a bulk, negative cosmological constant. Branes are placed at $\sigma = 0$ (+ve tension) and $\sigma = \pi r_c$ (-ve tension) – sourced by delta functions in the bulk energy momentum (the $|\sigma|$ in the warp factor ensures the delta function source terms). Thus, the bulk spacetime is a slice of AdS (between $\sigma = 0$ and $\sigma = \pi r_c$). Integrating the five dimensional Einstein–Hilbert action over the extra dimension one gets the relation between the four and five dimensional Planck scales (M_{Pl} and M, respectively),

$$M_{Pl}^2 = \frac{M^3}{k}\left(1 - e^{-2k\pi r_c}\right) \quad (2)$$

Therefore, for moderate $kr_c\pi$, the 4D Planck scale M_{Pl} is of the same order as the five dimensional Planck scale M, if $k \sim M_{Pl}$. On the other hand, writing down a massive wave equation in a scaled Minkowski metric (which is the metric on the brane hypersurface, see next section), one notices that physical masses (Higgs VeVs) on the visible brane at $\sigma = \pi r_c$ scale as

$$m = m_0 e^{-kr_c\pi} \quad (3)$$

where m_0 is the five dimensional mass-scale (say of the order of the Planck scale). m_0 is effectively warped by the factor $e^{-kr_c\pi}$ and the mass m seen on the brane can thus be at the TeV scale (assuming $kr_c \sim 12$). Hence, all fundamental scales (in five dimensions) are of the order of the Planck scale (no hierarchy) and the TeV scale on the brane arises because of extra dimensions and warping. This is the WED resolution of the hierarchy problem [6].

Journal of Physics: Conference Series **405** (2012) 012002 doi:10.1088/1742-6596/405/1/012002

1.3. Localisation
In the WED models, there is no notion of compactification (like in Kaluza Klein theory). The concept of localisation of fields on the brane hypersurface [7] replaces compactification. One assumes that all Standard Model fields are localised on the brane whereas gravity (and some other fields like the dilaton, Kalb Ramond, moduli) can spread into the bulk. However, to check localisation one needs to give the wave function representing the field an extra dimensional dependence and then prove that normalisable wave functions peaked about the location of the brane exist. For decaying warp factors, like the one for the RS model, the localisation problem reduces to that of a quantum particle in a volcano potential (see Fig. 2). Later in this article we will look at the problems that arise while analysing the localisation of gauge fields.

1.4. Thick branes
Apart from the RS type warp factor ($e^{-2k|\sigma|}$) wherein the brane is *thin* (delta function sources), there are models where the warp factor is a smooth function with no derivative discontinuities. These models are usually sourced by a scalar field in a potential. The brane in such *thick* brane models is realised as a scalar field domain wall (see [8, 9] for examples).

1.5. Stability
It is obvious that branes in the two-brane RS-type WED models are not stable. Stability is ensured by introducing an additional bulk scalar– this is the Goldberger–Wise mechanism [10]. An effective potential $V(r_c)$ is derived by integrating the matter (scalar field) action over the extra dimension. The existence of minima in the effective potential ensures stability. The $V(r_c)$ for the RS model does not have a minimum, but with additional scalar fields in the bulk, this becomes possible [10]. A useful discussion on stability may be found in [11].

1.6. Queries
The above basic formalism invites quite a few queries, some of which we now list below.

- Is a *scaled Minkowski metric*, which is the metric on the brane, *compatible* with all known laws of physics?
- Why is a *special* warped five dimensional line element chosen? Isn't a more *general* one possible? Say, for example:

$$ds^2 = -e^{2f_1(\sigma)}dt^2 + e^{2f_2(\sigma)}dx^2 + e^{2f_3(\sigma)}dy^2 + e^{2f_4(\sigma)}dz^2 + r_c^2 d\sigma^2 \tag{4}$$

- In the two brane model, can the branes have a *slant* w.r.t each other?
- The usual choice of *embedding* is $\sigma = \sigma_0$ (constant) and bulk (t, x, y, z) *identical* to on–brane (t, x, y, z). This is the *simplest* and is also *non–minimal*. Why this choice of embedding?

2. Lorentz violations
Let us now turn to the general issue in the first three queries mentioned above–that of Lorentz violations. We shall not discuss the fourth query in this article.

2.1. Scaled Minkowski
The line element on the visible brane at $\sigma = \pi r_c$ is given as

$$ds^2_{brane} = e^{-2kr_c\pi}\left[-dt^2 + dx^2 + dy^2 + dz^2\right]$$

This leads to a *violation of the principle of relativity*. Let us see why. Recall the *scaled* Lorentz transformations (see Pauli's book on Theory of Relativity [12]).

$$x' = \kappa\gamma\,(x - vt) \quad ; \quad y' = \kappa y \quad z' = \kappa z \quad ; \quad t' = \kappa\gamma\left(t - \frac{vx}{c^2}\right)$$

Thus, the distance function in the S' frame, written in the S frame coordinates is

$$ds'^2 = \kappa^2\left[-dt^2 + dx^2 + dy^2 + dz^2\right]$$

which is a scaled Minkowski line element. Note that the requirement $\kappa = 1$ follows from *reciprocity*, which, in turn, follows from the *principle of relativity*. Let us briefly recall what reciprocity is all about. Place a unit length rod in the S frame along y (the boost is along x, say). Its length in S' is κ. Now place the same rod in S'. Its length in S is $\frac{1}{\kappa}$. For reciprocity to hold, which is physically meaningful, we need $\kappa = 1$. Thus, with $\kappa \neq 1$ reciprocity and hence the principle of relativity are violated. We end up having a *preferred* frame. Thus, one needs to make a choice–*solve* heirarchy problem but *violate* the principle of relativity. Giving up the principle of relativity still *allows* some transformations between inertial frames, as demonstrated in a recent paper [13]. It may also be noted that von Ignatowsky in 1910 (see reference in [13]), first showed how the principle of relativity is *intimately tied* with the *group* structure of Lorentz transformations. With the *scaling* we have a larger group, the conformal group, and apart from Maxwell's equations in 4D other *massive* equations are *not* invariant, a fact which lies at the heart of the hierarchy resolution mentioned earlier. However, in the RS type WED models, the speed of light postulate of special relativity is not violated.

2.2. Asymmetric warping

We now move on towards more explicit Lorentz violations. This is achieved by the so–called asymmetrically warped line elements,

$$ds^2 = -e^{2f(\sigma)}dt^2 + e^{2g(\sigma)}|d\vec{r}|^2 + r_c^2 d\sigma^2$$

Thus, on the brane (at $\sigma = \sigma_0$), we get

$$ds^2 = -e^{2f(\sigma_0)}dt^2 + e^{2g(\sigma_0)}|d\vec{r}|^2$$

In such models [14] both the postulates of special relativity are violated. The speed of light can be greater than c, if $f(\sigma_0) \neq g(\sigma_0)$. This fact was exploited [15] to explain a recent claim on the existence of superluminal neutrinos, from a warped braneworld perspective. However, the experimental claim has now been found to be erroneous, but the theoretical construction in [15] remains an intriguing result.

2.3. Slanted branes

Another Lorentz violating scenario is the recently proposed slanted braneworld model [16]. The Hernandez–Sher tilted line element is given as

$$ds^2 = e^{-2C|\sigma| + ax|\sigma|}\eta_{ij}dx^i dx^j + \left(R^2 + Dxa\right)d\sigma^2$$

where a is the tilt parameter– a small quantity. The off–diagonal component of the Einstein tensor

$$G_{14} = -3\left(1 + \frac{CD}{2R^2}\right)a$$

which implies, for $G_{14} = 0$, $2R^2 = -CD$. The two branes are not parallel, but slightly tilted. Interestingly, to order a, almost all results of the RS two-brane model are retained. However, the tilt does affect the fermion masses through a spatially varying Higgs VeV. The electron–proton mass ratio is modified. Upper bounds on the tilt parameter a can be calculated using data from quasar spectra. These yield a value: $a < 7.5 \times 10^{-7} GPc^{-1}$. The tilt aspect is of course, meaningful in a two-brane model. It will be worth exploring this model in further detail– for example, its phenomenological and cosmological consequences.

3. Localisation again

Even though one can show that most of the physical fields are localisable on the brane in the RS type WED models, a problem arises with gauge fields. We shall now highlight this problem.

3.1. Gauge fields in RSI

In the two-brane model, the extra dimensional coordinate is bounded. However, it was shown in [17], for a $U(1)$ gauge theory, that the Kaluza Klein excitations of the gauge field are *strongly coupled* to fermions. Precision electroweak data constrain the *lowest* KK states to lie *above* 25 TeV. Taking the weak scale to be 1 TeV, the resulting implications on the model parameters force the bulk curvature to be larger than the higher dimensional Planck scale, M, a fact which violates theoretical consistency. To preserve $|R| < M^2$, the weak scale must be pushed to > 100 TeV.

A possible way out of the above problem is to introduce *brane-localised kinetic terms* [18] in the action, as follows,

$$S_g = -\frac{1}{4}\int d^5x \left\{\sqrt{-g_5}F_5^2 + \left[c_0\delta(y) + c_\pi\delta(y-\pi)\right]F_4^2\right\} \tag{5}$$

This modification leads to, for natural choices of parameters, a substantial *suppression* of the KK couplings, compared to the original model. Details may be found in [18]. However, though things work, note that such terms do not have a clear motivation in a curved spacetime setting and are introduced mainly as an extrapolation of flat spacetime results [18].

3.2. Gauge fields in RSII

In the single brane RS model, it is easy to see the localisation problem of gauge fields. Consider, the U(1) theory with equation of motion

$$\partial_i\left(\sqrt{-g}g^{ij}g^{kl}F_{jl}\right) = 0 \quad ; \quad \partial_i A^i = 0, A_5 = 0. \tag{6}$$

The equation for zero mode $A_i'' = 0$ yields a constant wave function along the fifth dimension $A_i = A_0 a_i(x)$. The constant wave function is normalisable for scalars but for Maxwell fields we have

$$\int d^5x\sqrt{-g}g^{ij}g^{kl}F_{ik}F_{jl} = \int dy A_0^2 \int d^4x F_{ij}^{(4)}F^{(4)ij} \tag{7}$$

and the y integration diverges. Therefore, free gauge fields cannot be localised in single brane models. A possible resolution is to modify the action as follows:

$$S_{gauge} = -\frac{1}{4}\int d^5x A(\phi) F_5^2$$

The $A(\phi)$ can be such that the divergent integral is rendered finite. Such modifications can arise due to various reasons: eg. dilaton, moduli couplings to gauge fields. Several other proposals also exist–however, the problem is far from resolved completely.

4. Brane gravity

There are three ways of studying brane gravity. We list them below and then discuss each briefly.

- Investigate the effects of the excited KK gravitons on particle phenomenology and find bounds on the masses of these states. This approach is ofcourse entirely perturbative.
- Look at the effective theory on the brane–the Shiromizu-Maeda-Sasaki equations [20] and focus on departures from standard Einstein gravity. A major problem here is the absence of a prescription to go back uniquely to the bulk.
- Investigate bulk higher dimensional gravity and its classical effects with reference to branes. This is a formal GR-motivated approach and needs to connect up with the other two.

4.1. Newtonian potential

The usual Newtonian gravitational potential is known to have corrections due to the presence of extra dimensions. The corrections due to the excited KK gravitons, have been calculated for both the LED and WED models. For instance, in the WED models, the corrected potential is of the form

$$U(r) = -\frac{Gm_1m_2}{r}\left(1+\frac{C}{(kr)^2}\right)$$

One might be able to test this correction in the ongoing Newtonian gravity experiments at EOTWASH and STANFORD. However, no such attempt has been made yet. The groups are primarily focused on Yukawa-type corrections which arise in LED models. Interestingly, a possible experiment involving the WED corrections has been suggested in [19] though it is yet to be realised.

4.2. Effective Einstein

What are the Einstein equations induced on the four dimensional brane from the five dimensional Einstein equations in the bulk? From the work of [20], we know that these equations are

$$G_{ij} = -\Lambda h_{ij} + \kappa^2 T_{ij} + 6\frac{\kappa^2}{\lambda}Q_{ij} - \mathcal{E}_{ij} \;\; ; \;\; \kappa^2 = \frac{1}{6}\lambda\kappa_5^4, \Lambda = \frac{1}{2}\left[\Lambda_5 + \kappa^2\lambda\right]$$
$$Q_{ij} = \frac{1}{12}TT_{ij} - \frac{1}{4}T_{ik}T_j^k + \frac{1}{24}h_{ij}\left[T_{kl}T^{kl} - T^2\right] \;\; ; \;\; \mathcal{E}_{ij} = {}^{(5)}C_{abcd}n^a n^c h_i^b h_j^d \tag{8}$$

The significant new terms here are a quadratic stress energy Q_{ij}, and the nonlocal $\mathcal{E}_{ij}$ which is traceless and has no off-brane component. The usual Einstein equations are recovered for situations where (i) the bulk Weyl is zero and (ii) the quadratic contributions are small. However, we can also show that the quadratic contributions are important in the early universe. Recently, similar quadratic contributions have been shown to arise in the Eddington–inspired Born–Infeld (EiBI) theory of gravity [22]. Though the quadratic stress energy terms in the EiBI theory are qualitatively similar there are quantitative differences. It would be interesting to see if there are any connections between the two.

The effects of $\mathcal{E}_{ij}$ have been explored in the context of black holes[21] and cosmology [21]. Further, a nonzero $\mathcal{E}_{ij}$ can be used to model dark matter[23]. But any nonzero $\mathcal{E}_{ij}$ would mean a non-zero bulk Weyl and therefore a bulk which is not of RS type. So, all conclusions using the RS metric will have to be revisited.

4.3. Bulk gravity

The study of bulk gravity is largely that of higher dimensional GR. One can begin with a generalised line element

$$ds^2 = e^{2f(\sigma)}\left(-dt^2 + a^2(t)|d\vec{x}|^2\right) + b^2(t)d\sigma^2$$

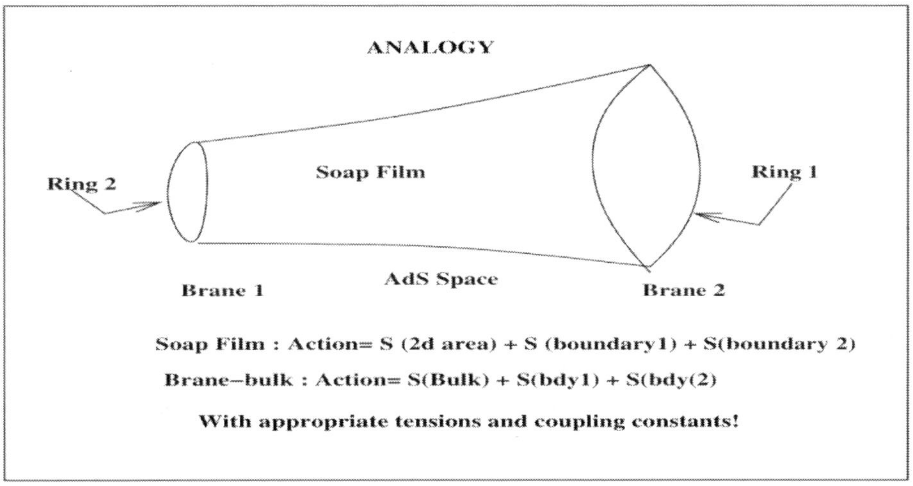

Figure 3. Lower dimensional analog model: soap film between rings.

and ask various questions. How does a variation of the warp factor reflect on geodesic motion? Is confined motion possible? How does a variation of the on-brane cosmological metric $a(t)$ and a time varying extra dimension $b(t)$, reflect on geodesic motion? What happens to geodesic flows (congruences) in the full five dimensional bulk? Do geodesic flows focus ? If so, when and how ? Under what conditions? Can we see/visualise the effect? What happens if there are more than one brane? Some of these questions are answered in [24].
Besides the above, there is signficant research on *Brane Cosmology*. We refrain from discussing it here and refer the reader to the excellent review in [21].

5. Analog model

A useful lower dimensional analog model for the warped two–brane system is that of a soap film suspended between two co-axial rings. Fig. 3 depicts the analogy and correspondence in detail. The geometry on the 2D surface is that of a negative curvature catenoid (minimal surface). One can study fluctuations about the minimal surface and it turns out that the effective potential for fluctuations is a volcano potential–similar to potentials that arise in the study of localisation of fields on the brane. The interested reader may look up [25] for further details on the soap film problem.

6. Bounds from LHC

Finally, let us turn to experiments and the latest bounds from LHC. Using diphoton events in 7 TeV proton-proton collisions with the ATLAS detector [26] the 95 % CL lower limits on the lightest RS graviton mass is between 0.79 and 1.85 TeV, for values of the dimensionless coupling $k/\bar{M}_{Pl}$ varying from 0.01 to 0.1. Combining with ATLAS results from the dielectron and dimuon final states, the 95% CL lower limit on the RS graviton mass for $k/\bar{M}_{Pl} = 0.01(0.1)$ is 0.80(1.95) TeV. Note, this is a lower bound and therefore not very conclusive.

7. Queries again and remarks

To end we provide a list of queries once again.

- What is there in the bulk? How do we decide on the content of bulk matter?
- How many extra dimensions are there? Is there a way to decide theoretically?

• What about models which consistently include the Lovelock, Gauss–Bonnet terms in higher dimensional gravity?

• Is it feasible to violate the principle of relativity in a classical world? How do we eliminate/retain asymmetrically warped or the slanted brane models?

• Is it possible to identify all SM particles using the localisation mechanism? Do we need a hybrid of compactification and localisation?

• How do we understand the quadratic stress energy and the Weyl–dependent terms better? What is their link with KK gravitons?

• How does one continue a four dimensional solution of the effective Einstein equations into the bulk? Taylor series type approaches give only local bulk geometry.

• Even if we have warped RS type extra dimensions, why will nature choose an AdS bulk? Unless experiments/observations decide on the existence of extra dimensions, one cannot answer many of the questions above. Even if experiments/observations prove the existence of extra dimensions, one would have to derive these phenomenological models from a fundamental theory, in a concrete way. Research in extra dimensions began in 1914 with Nordstrom's work. Maybe we will have good reasons to celebrate the centenary of extra dimensions, in 2014!

Acknowledgments
The author thanks the organisers of COSGRAV 2012 for their warm hospitality.

References

[1] Nordstrom G, 1914 *Phys. Zeit.* **15** 504; see also Ravndal F, gr-qc/0405030.
[2] Kaluza T, 1921 *Sitzungsber. Preuss. Akad. Wiss. Berlin. (Math. Phys.)*, 966; Klein O, 1926 *Zeitschrift fr Physik* **A 37 (12)** 895.
[3] Green M B, Schwarz J H and Witten E, 1987 *Superstring Theory, Vols I and II*, Cambridge University Press, Cambridge, UK.
[4] Akama K 1982, *"Pregeometry", in Lecture Notes in Physics, 176, Gauge Theory and Gravitation, Proceedings, Nara, edited by K. Kikkawa, N. Nakanishi and H. Nariai, (Springer-Verlag)*, 267; Rubakov V A and Shaposhnikov M E 1983, *Phys. Lett.* **125B** 136; Visser M 1985, *Phys.Lett.* **B159** 22 ; Gogberashvili M, hep-ph/9812296, hep-ph/9812365, hep-ph/9904383, hep-ph/9908347.
[5] Arkani-Hamed N, Dimopoulos S, Dvali G 1998 *Phys. Letts.* **B429** 263; Antoniadis I, Arkani-Hamed N, Dimopoulos S, Dvali G 1998 *Phys. Letts.* **B 436** 257.
[6] Randall L, Sundrum R, 1999 *Phys. Rev. Letts.* **83** 3370; *ibid.* **83** 4690.
[7] Bajc B and Gabadadze G, 2000 *Phys. Letts.* **B474** 282; Randjbar-Daemi S and Shaposhnikov M, 2000 *Phys. Letts.* **492** 361.
[8] Csaki C, arXiv:hep-ph/0404096.
[9] Koley R and Kar S, 2005 *Class. Qtm. Grav.* **22** 753.
[10] Goldberger A and Wise M, 1999 *Phys. Rev. Letts.* **83** 4922.
[11] Das A, Kar S and SenGupta S 2009 *Int. Jr. Mod. Phys.* **A24** 4457.
[12] Pauli W, 1981 *Theory of Relativity*, Dover Publications.
[13] Baccetti V, Tate K and Visser M, arXiv:1112.1466.
[14] Csaki C, Ehrlich J and Grojean C 2001, *Nucl. Phys.* **B 312** 312.
[15] Gubser S S arXiv:1109.5687.
[16] Hernandez D and Sher M, 2011 *Phys.Lett.* **B698** 403.
[17] Davoudiasl H, Hewett J L and Rizzo T G, 2000 *Phys.Lett.* **B473** 43.
[18] Davoudiasl H, Hewett J L and Rizzo T G, 2003 *Phys.Rev.* **D68** 045002.
[19] Azam M, Sami M, Unnikrishnan C S and Shiromizu T, 2008 *Phys. Rev.* **D77** 101101.
[20] Shiromizu T, Maeda K and Sasaki M, 2000 *Phys. Rev.* **D 62** 024012.
[21] Maartens R, 2010 *Living Rev. Relativity* **13** 5.
[22] Banados M and Ferreira P G 2010 *Phys. Rev. Letts.* **105** 011101.
[23] Pal S, Bharadwaj S and Kar S 2005 *Phys. Letts.* **B609** 194.
[24] Ghosh S , Nandan H and Kar S 2010 *Phys. Rev.* **D82** 024040; Ghosh S , DasGupta A and Kar S 2011 *Phys. Rev.* **D83** 084001.
[25] Durand L 1981 *Am. Jr. Phys* **49** 334.
[26] The ATLAS Collaboration, arxiv: 1112.2194.

Non-Gaussianities from isocurvature modes

David Langlois
APC, Astroparticules et Cosmologie, Université Paris Diderot, CNRS/IN2P3, CEA/Irfu, Observatoire de Paris, Sorbonne Paris Cité, 10, rue Alice Domon et Léonie Duquet, 75205 Paris Cedex 13, France

E-mail: langlois@apc.univ-paris7.fr

Abstract. This contribution discusses isocurvature modes, in particular the non-Gaussianities of local type generated by these modes. Since the isocurvature transfer functions differ from the adiabatic one, the coexistence of a primordial isocurvature mode with the usual adiabatic mode leads to a rich structure of the angular bispectrum, which can be decomposed into six elementary bispectra. Future analysis of the CMB data will enable to measure their relative weights, or at least constrain them. Non-Gaussianity thus provides a new window on isocurvature modes. This is particularly relevant for some scenarios, such as those presented here, which generate isocurvature modes whose contribution in the power spectrum is suppressed, as required by present data, but whose contribution in the non-Gaussianities could be dominant and measurable.

1. Introduction

Inflation is currently the best candidate to explain the generation of primordial perturbations, but current observations cannot point to a specific scenario. The hope is thus that future data will enable us to discriminate between various models. In this respect, an important distinction is between single-field and multiple-field models. The detection of even a tiny fraction of isocurvature mode in the cosmological data would rule out single-field inflation, which predicts only adiabatic perturbations. Another signature that could distinguish multiple-field models from single-field models is a detectable primordial non-Gaussianity of the local type. Similarly to isocurvature modes, a detection of local primordial non-Gaussianity would rule out all inflation models based on a single scalar field, since they generate only unobservably small local non-Gaussianities.

This contribution, based on the recent works [1, 2, 3, 4], discusses how isocurvature modes could affect non-Gaussianities and their specific signature in the Cosmic Microwave Background (CMB) fluctuations. In particular, it is shown how the bispectrum generated by the adiabatic mode together with one isocurvature mode leads to a total angular bispectrum which can be decomposed into six distinct components: the usual purely adiabatic bispectrum, a purely isocurvature bispectrum, and four other bispectra that arise from the possible correlations between the adiabatic and isocurvature mode. Because these six bispectra have different shapes in angular space, their amplitude can in principle be measured in the CMB data.

This analysis opens a new window on isocurvature modes, especially important to test models where the isocurvature contribution is suppressed in the power spectrum but not in the non-Gaussianities. Examples of such models are presented at the end of this contribution.

Published under licence by IOP Publishing Ltd

2. Isocurvature modes

At the time of last scattering, the main components in the Universe are the CDM (c), the baryons (b), the photons (γ) and the neutrinos (ν). All these components are characterized by their individual energy density contrasts δ_i. The most common type of perturbation is the adiabatic mode, characterized by the condition

$$\delta_c = \delta_b = \frac{3}{4}\delta_\nu = \frac{3}{4}\delta_\gamma \,, \tag{1}$$

which means that the number of photons (or neutrinos, or CDM particles) per baryon does not fluctuate. Assuming adiabatic initial conditions is natural if all particles have been created by the decay of a single degree of freedom, such as a single inflaton, and, so far, the CMB data are fully compatible with purely adiabatic perturbations.

However, other types of perturbations can be included in a more general framework. In addition to the adiabatic mode, one can consider four distinct isocurvature modes [5]: the CDM isocurvature mode, the baryon isocurvature mode, the neutrino density isocurvature mode and the neutrino velocity isocurvature mode. The first three isocurvature modes are characterized by

$$S_X = \frac{1}{1+w_X}\delta_X - \frac{3}{4}\delta_\gamma, \qquad X = \{c, b, \nu d\} \tag{2}$$

with $w_X = P_X/\rho_X$. As for the neutrino velocity isocurvature mode, it is characterized by a non vanishing "initial velocity", compensated by the velocity of the photon-baryon plasma so that the total momentum density is cancelled, while the energy densities satisfy (1).

In the following, these five modes will be denoted collectively as X^I. In the context of inflation, a necessary, although not sufficient, condition for at least one of these isocurvature modes to be produced is that several light degrees of freedom exist during inflation. Moreover, since the adiabatic and isocurvature modes can be related in various ways to these degrees of freedom during inflation, one can envisage the existence of correlations between these modes [6].

The various modes lead to *different* predictions for the CMB temperature and polarization. Let us consider for instance the temperature anisotropies, which can be decomposed into spherical harmonics:

$$\frac{\Delta T}{T} = \sum_{lm} a_{lm} Y_{lm} \,. \tag{3}$$

At linear order, the multipole coefficients a_{lm} are related to a general primordial perturbation consisting of the superposition of several modes, via the expression

$$a_{lm} = 4\pi(-i)^l \int \frac{d^3\mathbf{k}}{(2\pi)^3}\left(\sum_I X^I(\mathbf{k})\, g_l^I(k)\right) Y_{lm}^*(\hat{\mathbf{k}}) \,, \tag{4}$$

where $g_l^I(k)$ denotes the transfer function associated with the mode X^I. As a result, the total angular power spectrum is given by

$$C_l = \langle a_{lm} a_{lm}^* \rangle = \sum_{I,J} \frac{2}{\pi} \int_0^\infty dk\, k^2 g_l^I(k) g_l^J(k) P_{IJ}(k) \,, \tag{5}$$

where the primordial power spectra $P_{IJ}(k)$ are defined by

$$\langle X^I(\mathbf{k}_1) X^J(\mathbf{k}_2)\rangle \equiv (2\pi)^3 \delta(\mathbf{k}_1 + \mathbf{k}_2) P_{IJ}(k_1) \,. \tag{6}$$

Since the various transfer functions are *different*, this leads to different predictions for the CMB angular power spectrum. This is illustrated in Fig. 1 (left panel), where the angular power

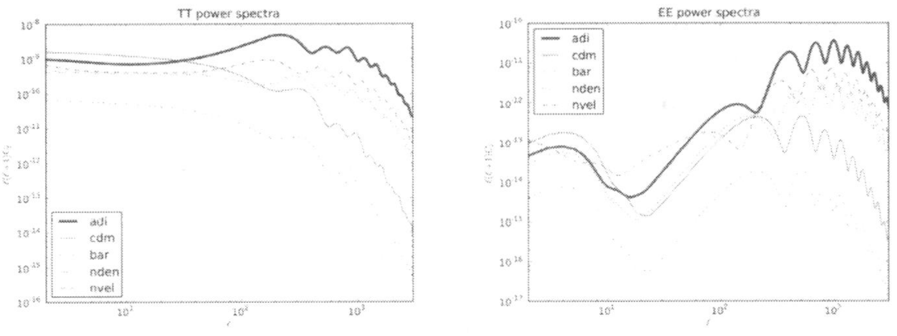

Figure 1. Angular power spectra (multiplied by $l(l+1)$) for the temperature (left) and polarization (right) obtained from purely adiabatic or purely isocurvature initial conditions. The amplitude and spectral index of the primordial power spectrum, as well as the cosmological parameters, on which the transfer functions depend, correspond to the WMAP7-only best-fit parameters.

spectra produced separately by the various modes are plotted, assuming the same primordial power spectrum. The only exception are the CDM and baryon isocurvature modes which give exactly the same pattern, up to the rescaling $S_b = (\Omega_b/\Omega_c)\, S_c$ where Ω_b and Ω_c denote, as usual, the present energy density fractions, respectively for baryons and CDM.

We infer from CMB observations that the "primordial" perturbation is mainly of the adiabatic type. However, this does not preclude the presence, in addition to the adiabatic mode, of an isocurvature component, with a smaller amplitude. Precise measurement of the CMB fluctuations could lead to a detection of such an extra component, or at least put constraints on its amplitude. For example, constraints on the CDM isocurvature to adiabatic ratio,

$$\alpha = \frac{\mathcal{P}_{S_c}}{\mathcal{P}_\zeta}, \tag{7}$$

based on the WMAP7+BAO+SN data, have been published for the uncorrelated and fully correlated cases (the impact of isocurvature perturbations on the observable power spectrum indeed depends on the correlation between adiabatic and isocurvature perturbations, as illustrated in [7]). In terms of the parameter $a \equiv \alpha/(1+\alpha)$, the limits given in [8] are

$$a_0 < 0.064 \quad (95\%\text{CL}), \qquad a_1 < 0.0037 \quad (95\%\text{CL})\,, \tag{8}$$

respectively for the uncorrelated case and for the fully correlated case.

3. Generalized angular bispectra

Let us now turn to the CMB non-Gaussianities. The angular bispectrum corresponds to the three-point function of the multipole coefficients:

$$B_{l_1 l_2 l_3}^{m_1 m_2 m_3} \equiv \langle a_{l_1 m_1} a_{l_2 m_2} a_{l_3 m_3} \rangle\,. \tag{9}$$

Substituting the expression (4) into the angular bispectrum, one finds

$$B_{l_1 l_2 l_3}^{m_1 m_2 m_3} = \mathcal{G}_{l_1 l_2 l_3}^{m_1 m_2 m_3} b_{l_1 l_2 l_3}\,, \tag{10}$$

where the first, purely geometrical, factor is the Gaunt integral

$$\mathcal{G}_{l_1 l_2 l_3}^{m_1 m_2 m_3} \equiv \int d^2\hat{\mathbf{n}}\, Y_{l_1 m_1}(\hat{\mathbf{n}})\, Y_{l_2 m_2}(\hat{\mathbf{n}})\, Y_{l_3 m_3}(\hat{\mathbf{n}})\,, \tag{11}$$

while the second factor, usually called the *reduced* bispectrum,

$$b_{l_1l_2l_3} = \sum_{I,J,K}\left(\frac{2}{\pi}\right)^3\int\left(\prod_{i=1}^{3}k_i^2dk_i\right)\, g_{l_1}^I(k_1)g_{l_2}^J(k_2)g_{l_3}^K(k_3) \times B^{IJK}(k_1,k_2,k_3)\int_0^\infty r^2dr\,j_{l_1}(k_1r)j_{l_2}(k_2r)j_{l_3}(k_3r)\,, \quad (12)$$

depends on the bispectra of the primordial X^I:

$$\langle X^I(\mathbf{k}_1)X^J(\mathbf{k}_2)X^K(\mathbf{k}_3)\rangle \equiv (2\pi)^3\delta(\Sigma_i\mathbf{k}_i)B^{IJK}(k_1,k_2,k_3)\,. \quad (13)$$

We now need to specify the primordial bispectra B^{IJK}. For purely adiabatic perturbations, the local bispectrum is expressed as the square of the power spectrum (symmetrized over k_1, k_2 and k_3). Here, we consider the generalization

$$B^{IJK}(k_1,k_2,k_3) = \tilde{f}_{\rm NL}^{I,JK}P_\zeta(k_2)P_\zeta(k_3) + \tilde{f}_{\rm NL}^{J,KI}P_\zeta(k_1)P_\zeta(k_3) + \tilde{f}_{\rm NL}^{K,IJ}P_\zeta(k_1)P_\zeta(k_2)\,, \quad (14)$$

where the coefficients $\tilde{f}_{\rm NL}^{I,JK}$ satisfy the condition

$$\tilde{f}_{\rm NL}^{I,JK} = \tilde{f}_{\rm NL}^{I,KJ}\,. \quad (15)$$

The above expression is the natural outcome of a generic model of multiple-field inflation. Indeed, allowing for several light degrees of freedom during inflation, one can relate, in a very generic way, the "primordial" perturbations X^I (defined during the standard radiation era) to the fluctuations of light primordial fields ϕ^a, generated at Hubble crossing during inflation, so that one can write, up to second order,

$$X^I = N_a^I\,\delta\phi^a + \frac{1}{2}N_{ab}^I\,\delta\phi^a\delta\phi^b + \ldots \quad (16)$$

where the $\delta\phi^a$ can usually be treated as independent quasi-Gaussian fluctuations, i.e.

$$\langle\delta\phi^a(\mathbf{k})\,\delta\phi^b(\mathbf{k}')\rangle = (2\pi)^3\,\delta^{ab}P_{\delta\phi}(k)\,\delta(\mathbf{k}+\mathbf{k}')\,, \qquad P_{\delta\phi}(k) = 2\pi^2k^{-3}\left(\frac{H_*}{2\pi}\right)^2\,, \quad (17)$$

where a star denotes Hubble crossing time. The relation (16) is very general, and all the details of the inflationary model are embodied by the coefficients N_a^I and N_{ab}^I. Substituting (16) into (13) and using Wick's theorem, one finds that the bispectra B_{IJK} can be expressed in the form

$$B^{IJK}(k_1,k_2,k_3) = \lambda^{I,JK}P_{\delta\phi}(k_2)P_{\delta\phi}(k_3) + \lambda^{J,KI}P_{\delta\phi}(k_1)P_{\delta\phi}(k_3) + \lambda^{K,IJ}P_{\delta\phi}(k_1)P_{\delta\phi}(k_2)\,, \quad (18)$$

with the coefficients

$$\lambda^{I,JK} \equiv \delta^{ac}\delta^{bd}N_{ab}^IN_c^JN_d^K \quad (19)$$

(the summation over scalar field indices a, b, c and d is implicit), which are symmetric under the interchange of the last two indices, by construction. Since the adiabatic power spectrum is given by

$$P_\zeta = (\delta^{ab}N_a^\zeta N_b^\zeta)P_{\delta\phi} \equiv AP_{\delta\phi}, \quad (20)$$

one obtains finally (14) with

$$\tilde{f}_{\rm NL}^{I,JK} = \lambda_{NL}^{I,JK}/A^2\,, \quad (21)$$

where it is implicitly assumed that the coefficients N_a^I are weakly time dependent so that the scale dependence of A^2 can be neglected.

After substitution of (14) into (12), the reduced bispectrum can finally be written as

$$b_{l_1 l_2 l_3} = \sum_{I,J,K} \tilde{f}_{\rm NL}^{I,JK} b_{l_1 l_2 l_3}^{I,JK}, \tag{22}$$

where each contribution is of the form[1]

$$b_{l_1 l_2 l_3}^{I,JK} = 3 \int_0^\infty r^2 dr\, \alpha_{(l_1}^I(r) \beta_{l_2}^J(r) \beta_{l_3)}^K(r), \tag{23}$$

with

$$\alpha_l^I(r) \equiv \frac{2}{\pi} \int k^2 dk\, j_l(kr)\, g_l^I(k), \qquad \beta_l^I(r) \equiv \frac{2}{\pi} \int k^2 dk\, j_l(kr)\, g_l^I(k)\, P_\zeta(k)\,. \tag{24}$$

4. Observational prospects

For simplicity, we assume that the primordial perturbation is the combination of the dominant adiabatic mode with a *single* isocurvature mode. In this case, the total bispectrum is characterized by *six* parameters, which we now denote $\tilde{f}^{(i)}$,

$$\begin{aligned} b_{l_1 l_2 l_3} &= \tilde{f}^{\zeta,\zeta\zeta} b_{l_1 l_2 l_3}^{\zeta,\zeta\zeta} + 2\tilde{f}^{\zeta,\zeta S} b_{l_1 l_2 l_3}^{\zeta,\zeta S} + \tilde{f}^{\zeta,SS} b_{l_1 l_2 l_3}^{\zeta,SS} + \tilde{f}^{S,\zeta\zeta} b_{l_1 l_2 l_3}^{S,\zeta\zeta} + 2\tilde{f}^{S,\zeta S} b_{l_1 l_2 l_3}^{S,\zeta S} + \tilde{f}^{S,SS} b_{l_1 l_2 l_3}^{S,SS} \\ &= \sum_{(i)} \tilde{f}^{(i)} b_{l_1 l_2 l_3}^{(i)}, \end{aligned} \tag{25}$$

where the index i varies between 1 to 6, following the order indicated in the upper line. Note that, because of the factor 2 in front of $\tilde{f}^{\zeta,\zeta S}$ and $\tilde{f}^{S,\zeta S}$, we define $b_{l_1 l_2 l_3}^{(2)} \equiv 2 b_{l_1 l_2 l_3}^{\zeta,\zeta S}$ and $b_{l_1 l_2 l_3}^{(5)} \equiv 2 b_{l_1 l_2 l_3}^{S,\zeta S}$ whereas there is no such factor 2 for the other terms.

To estimate these six parameters, given some data set, the usual procedure is to minimize

$$\chi^2 = \left\langle (B^{obs} - \sum_i \tilde{f}^{(i)} B^{(i)}), (B^{obs} - \sum_i \tilde{f}^{(i)} B^{(i)}) \right\rangle, \tag{26}$$

where B is a short notation for the angle-averaged bispectrum

$$B_{l_1 l_2 l_3} \equiv \sum_{m_1, m_2, m_3} \begin{pmatrix} l_1 & l_2 & l_3 \\ m_1 & m_2 & m_3 \end{pmatrix} B_{l_1 l_2 l_3}^{m_1 m_2 m_3}\,. \tag{27}$$

For an ideal experiment (no noise and no effects due to the beam size) without polarization, the scalar product in (26) is defined by

$$\langle B, B' \rangle \equiv \sum_{l_1 \le l_2 \le l_3} \frac{B_{l_1 l_2 l_3} B'_{l_1 l_2 l_3}}{\sigma_{l_1 l_2 l_3}^2}. \tag{28}$$

with the variance

$$\sigma_{l_1 l_2 l_3}^2 \equiv \langle B_{l_1 l_2 l_3}^2 \rangle - \langle B_{l_1 l_2 l_3} \rangle^2 \approx (1 + \delta_{l_1 l_2} + \delta_{l_2 l_3} + \delta_{l_3 l_1} + 2\,\delta_{l_1 l_2} \delta_{l_2 l_3})\, C_{l_1} C_{l_2} C_{l_3} \tag{29}$$

in the approximation of weak non-Gaussianity.

The best estimates for the parameters are thus obtained by solving

$$\sum_j \langle B^{(i)}, B^{(j)} \rangle \tilde{f}^{(j)} = \langle B^{(i)}, B^{obs} \rangle\,, \tag{30}$$

[1] We use the standard notation: $(l_1 l_2 l_3) \equiv [l_1 l_2 l_3 + 5\,\text{perms}]/3!$.

while the statistical error on the parameters is deduced from the second-order derivatives of χ^2, which define the Fisher matrix, given here by

$$F_{ij} \equiv \langle B^{(i)}, B^{(j)} \rangle. \tag{31}$$

For a real experiment, and if E-polarization is included as well, the above equations remain valid, except that the definition of the scalar product has to be replaced by a more complicated expression (see [4] for details).

For each of the four isocurvature modes, the corresponding Fisher matrix has been computed numerically, including the polarization, for the noise characteristics of the Planck satellite in [3, 4]. The error on the parameters $\tilde{f}^i$ can then be deduced from the components of the Fisher matrix, via the expression

$$\Delta \tilde{f}^i = \sqrt{(F^{-1})_{ii}} \tag{32}$$

For the various cases, we have obtained

$$\Delta \tilde{f}^i = \{9.6, 7.1, 160, 150, 180, 140\} \quad \text{(CDM isocurvature)} \tag{33}$$
$$\Delta \tilde{f}^i = \{9.6, 35, 4000, 720, 4300, 16600\} \quad \text{(baryon isocurvature)} \tag{34}$$
$$\Delta \tilde{f}^i = \{28, 36, 190, 150, 240, 320\} \quad \text{(neutrino density isocurvature)} \tag{35}$$
$$\Delta \tilde{f}^i = \{25, 22, 85, 81, 77, 71\} \quad \text{(neutrino velocity isocurvature)} \tag{36}$$

As one can see, the error on the first two parameters is much smaller than the last four parameters in the CDM isocurvature case. An explanation for this result is given in [4].

5. Illustrative example

To illustrate the previous results, which are model-independent, it is instructive to consider a simple class of models based on the presence of a spectator light scalar field during inflation, dubbed curvaton. This curvaton acquires nearly scale-invariant fluctuations during inflation and, later, behaves as a pressureless fluid when it oscillates at the bottom of its potential, before decaying.

Here, we allow the curvaton σ to decay into both radiation and CDM with the respective branching ratios γ_r and γ_c. Since, in general, CDM can already be present before the decay, we define the fraction of CDM created by the decay as $f_c \equiv \gamma_c \Omega_\sigma/(\Omega_c + \gamma_c \Omega_\sigma)$, where the Ω's represent the relative abundances just before the decay.

As shown in [1], the "primordial" adiabatic and isocurvature perturbations, i.e. defined after the curvaton decay, can be written in the form (16), with

$$N_\sigma^\zeta = \frac{2r}{3\sigma_*}, \qquad N_{\sigma\sigma}^\zeta = \frac{2r}{3\sigma_*^2}, \tag{37}$$

$$N_\sigma^S = \frac{2}{\sigma_*}(f_c - r), \qquad N_{\sigma\sigma}^S = \frac{2}{\sigma_*^2}\left[f_c(1-2f_c) - r\right], \tag{38}$$

where $r \equiv 3\gamma_r \Omega_\sigma/[(4-\Omega_\sigma)(1-(1-\gamma_r)\Omega_\sigma)]$ is assumed to be small, since significant non-Gaussianities arise only if $r \ll 1$.

Let us first discuss linear perturbations. It is useful to introduce the curvaton contribution to the total adiabatic power spectrum $\Xi \equiv (N_\sigma^\zeta)^2/[(N_\phi^\zeta)^2 + (N_\sigma^\zeta)^2]$, where $N_\phi^\zeta = H/\dot{\phi}$ is associated with the inflaton fluctuation, and $N_\phi^S = 0$. Ξ is directly related to the correlation $\mathcal{C} \equiv P_{\zeta,S}/\sqrt{P_S P_\zeta} = \sqrt{\Xi}\,\mathrm{sgn}(f_c - r)$. The isocurvature-adiabatic ratio, given by

$$\alpha \equiv \frac{P_S}{P_\zeta} = \frac{(N_\sigma^S)^2}{(N_\phi^\zeta)^2 + (N_\sigma^\zeta)^2} = 9\left(1 - \frac{f_c}{r}\right)^2 \Xi, \tag{39}$$

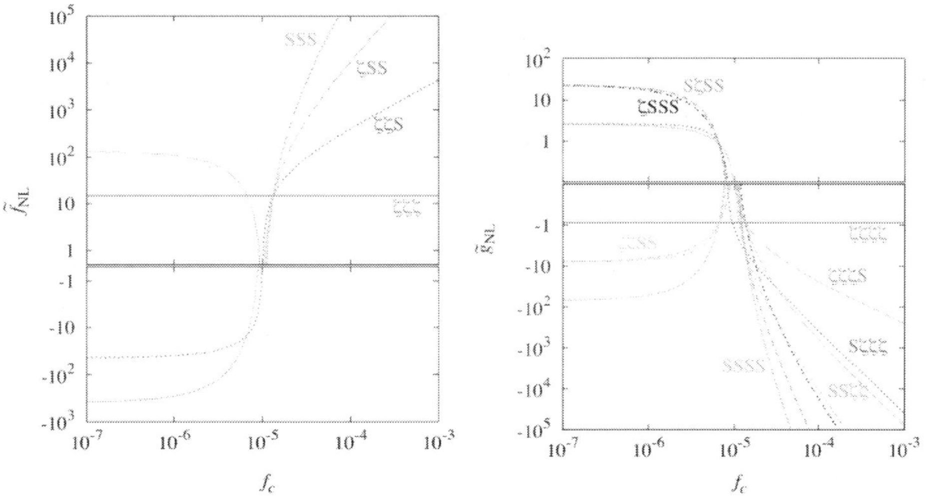

Figure 2. Plots of the coefficients $\tilde{f}_{\rm NL}^{I,JK}$ (left panel) and of $\tilde{g}_{\rm NL}^{IJKL}$ (right panel) as functions of f_c. I, J, K, L are specified in the figure for each line. Other parameters are fixed as $\xi = 1, \lambda = 10^{-3}$ and $r = 10^{-5}$.

is constrained by CMB observations [8] to be small which requires either $f_c \simeq r$ or $\Xi \ll 1$.

Let us now turn to non-Gaussianities. Using (37), one finds $\tilde{f}_{\rm NL}^{\zeta,\zeta\zeta} = 3\,\Xi^2/(2r)$. This is the dominant contribution in the regime $f_c \simeq r$, the other components being suppressed. We thus concentrate on the more interesting case $\Xi \ll 1$ to discuss the size of the various components in terms of f_c and r, considered as free parameters in our phenomenological approach.

In the regime $f_c \ll r \ll 1$, the purely adiabatic coefficient is the smallest one. The other ones are *enhanced* by powers of (-3) (since $N^S_\sigma/N^\zeta_\sigma = N^S_{\sigma\sigma}/N^\zeta_{\sigma\sigma} = -3$):

$$\tilde{f}_{\rm NL}^{I,JK} = (-3)^p \tilde{f}_{\rm NL}^{\zeta,\zeta\zeta}, \quad \tilde{f}_{\rm NL}^{\zeta,\zeta\zeta} = \frac{\alpha^2}{54r}, \tag{40}$$

where p is the number of "S" in the triplet $\{I, J, K\}$. In particular, the purely isocurvature coefficient is enhanced by a factor 27, but with the opposite sign: $\tilde{f}_{\rm NL}^{S,SS} = -\alpha^2/(2r)$. All coefficients can be significant if r is sufficiently smaller than α^2.

In the opposite regime $r \ll f_c \ll 1$, the purely adiabatic coefficient is, once again, the smallest one. All the coefficients are now positive and enhanced by factors $(3f_c/r)^p$, where p is again the number of "S" indices:

$$\tilde{f}_{\rm NL}^{I,JK} = \left(\frac{3f_c}{r}\right)^p \tilde{f}_{\rm NL}^{\zeta,\zeta\zeta}, \quad \tilde{f}_{\rm NL}^{\zeta,\zeta\zeta} = \frac{\alpha^2 r^3}{54 f_c^4}. \tag{41}$$

Note that the enhancement factor is much bigger than in the previous case (40). The purely isocurvature coefficient, which dominates, is $\tilde{f}_{\rm NL}^{S,SS} = \alpha^2/(2f_c)$ and can be large if f_c is sufficiently small, while the relative size of the other coefficients depends on the ratio r/f_c. The full dependence of the coefficients on the parameter f_c is illustrated in the left panel of Fig. 2.

A similar analysis for the trispectrum has been presented in [2]. The usual coefficients $\tau_{\rm NL}$ and $g_{\rm NL}$ that describe the local trispectrum are then generalized into, respectively, nine $\tau_{\rm NL}$-like coefficients and eight $g_{\rm NL}$-like coefficients:

$$\tau_{\rm NL}^{IJ,KL}, \qquad \tilde{g}_{\rm NL}^{I,JKL} \equiv \frac{54}{25} g_{\rm NL}^{I,JKL}. \tag{42}$$

The behaviour of the $g_{\rm NL}$-like coefficients as functions of f_c is plotted on the right panel of Fig. 2. The hierarchies are very similar to those observed for the bispectrum parameters.

In conclusion, the above results show that a small isocurvature fraction in the power spectrum is compatible with a dominantly isocurvature non-Gaussianity detectable by Planck (e.g. $\alpha \simeq 10^{-2}$ and $r \ll f_c \simeq 10^{-8}$ yields $\tilde{f}_{\rm NL}^{S,SS} \simeq 5 \times 10^3$). Of course, the relations (40) or (41), are specific to the models considered here and would be a priori different in other models. It is therefore important to try to measure these six coefficients *separately*, in order to obtain model-independent constraints from observations.

Acknowledgments

I would like to thank the organizers of the COSGRAV12 conference for their warm hospitality in Kolkata. I am also grateful to A. Lepidi, T. Takahashi, B. van Tent for their contribution to the results presented here.

References

[1] D. Langlois, A. Lepidi, JCAP **1101**, 008 (2011). [arXiv:1007.5498 [astro-ph.CO]].
[2] D. Langlois, T. Takahashi, JCAP **1102**, 020 (2011). [arXiv:1012.4885 [astro-ph.CO]].
[3] D. Langlois and B. van Tent, Class. Quant. Grav. **28**, 222001 (2011) [arXiv:1104.2567 [astro-ph.CO]].
[4] D. Langlois and B. van Tent, "Isocurvature Modes in the CMB bispectrum," arXiv:1204.5042 [astro-ph.CO].
[5] M. Bucher, K. Moodley, N. Turok, Phys. Rev. **D62**, 083508 (2000). [astro-ph/9904231].
[6] D. Langlois, Phys. Rev. **D59**, 123512 (1999). [astro-ph/9906080].
[7] D. Langlois, A. Riazuelo, Phys. Rev. **D62**, 043504 (2000). [astro-ph/9912497].
[8] E. Komatsu *et al.* [WMAP Collaboration], Astrophys. J. Suppl. **192**, 18 (2011) [arXiv:1001.4538 [astro-ph.CO]].

International Conference on Modern Perspectives of Cosmology and Gravitation (COSGRAV12) IOP Publishing
Journal of Physics: Conference Series **405** (2012) 012004 doi:10.1088/1742-6596/405/1/012004

The Quantum State of Inflationary Perturbations

Jérôme Martin

Institut d'Astrophysique de Paris, UMR 7095-CNRS, Université Pierre et Marie Curie, 98 bis boulevard Arago, 75014 Paris, France

E-mail: jmartin@iap.fr

Abstract. This article reviews the properties of the quantum state of inflationary perturbations. After a brief description of the inflationary background, the wavefunction of the Mukhanov-Sasaki variable is calculated and shown to be that of a strongly squeezed state. The corresponding Wigner function and density matrix, which are convenient tools to characterize the properties of a quantum state, are also evaluated. Finally, the issue of definite outcomes for inflation is discussed.

1. Introduction

According to the theory of inflation, the very early universe underwent a period of quasi-exponential expansion [1, 2, 3, 4, 5]. Such a phase of evolution can arise in general relativity when the pressure of the fluid sourcing the Einstein equations is negative. This is the case if a scalar field with a flat potential dominated in the very early universe. This simple assumption, which solely rests on classical physics, is quite remarkable as it allows us to solve many puzzles of the former standard model of cosmology, *i.e.* the hot Big Bang scenario.

But even more interesting results can be obtained when quantum effects are taken into account. In particular, considering small inhomogeneous fluctuations on top of an isotropic and homogeneous expanding background and quantizing them gives a very elegant explanation for the origin of the Cosmic Microwave Background (CMB) anisotropies and of the large scale structures [6, 7, 8, 9, 10, 11] (for reviews, see Refs. [12, 13, 14, 15, 16, 17, 18]). As is well-known, this leads to an almost scale-invariant power spectrum for the cosmological perturbations [19], the small deviations from scale invariance being connected to the detailed shape of the scalar field potential. As a consequence, this provides a method to constrain the physical conditions that prevailed when the universe was very young, at energy scales that can be as high as 10^{15}GeV [20, 21, 22, 23, 24, 25, 26].

This is not, however, the only interesting aspect that arises when quantum effects are included into the inflationary picture. Very fundamental questions, connected to the quantum nature of the gravitational field and/or to the foundations of quantum mechanics, can also be discussed in this framework. In these proceedings, we review the properties of the quantum state of inflationary perturbations [27, 28] and briefly discuss how it is related to the more fundamental issues mentioned above [27, 28, 29, 30, 31, 32, 33, 34, 35, 36, 37, 38]. This short article is organized as follows. In the next section, Sec. 2, we describe inflation at the background level. Then, in Sec. 3, we review the theory of inflationary fluctuations of quantum-mechanical origin, paying special attention to the quantum state in which the perturbations are placed. Finally, in Sec. 4, we present our conclusions and briefly mention and discuss the open issues.

Journal of Physics: Conference Series **405** (2012) 012004 doi:10.1088/1742-6596/405/1/012004

2. The Inflationary Background

Inflation is a phase of accelerated, quasi-exponential, expansion that took place in the very early Universe, prior to the standard hot Big-Bang phase [1, 2, 3, 4, 5] (for reviews, see Refs. [13, 14, 15]). It is important to realize that inflation does not replace the former cosmological standard model by a new one but just completes it. Postulating such a phase of evolution allows us to solve the standard problems of the hot Big-Bang model. Given that, at very high energies, field theory is the relevant framework to describe matter, a natural way to realize inflation is to consider that a real scalar field (the "inflaton" field) was the dominant source of energy density in the early Universe (there also exist versions of the inflationary scenario where inflation is driven by several scalar fields). This assumption is moreover compatible with the observed homogeneity, isotropy and flatness of the early Universe. Technically, this can be described by the metric tensor $ds^2 = -dt^2 + a^2(t)\delta_{ij}dx^i dx^j$, where $a(t)$ is the Friedman-Lemaître-Robertson-Walker (FLRW) scale factor and t the cosmic time. The Einstein equations imply that $\ddot{a}/a = -(\rho+3p)/(6M_{\rm Pl}^2)$, ρ and p being the energy density and pressure of the matter sourcing the gravitational field and $M_{\rm Pl}$ the Planck mass (a dot denotes a derivative with respect to the cosmic time t). For a scalar field, this reduces to $\ddot{a}/a = V(\varphi)(1-\dot{\varphi}^2/V)/(3M_{\rm Pl}^2)$, where $V(\varphi)$ is the scalar field potential. This means that inflation (*i.e.* $\ddot{a} > 0$) can be obtained provided the inflaton slowly rolls down its potential so that its potential energy dominates over its kinetic energy. This also shows that the inflaton potential must be sufficiently flat, a requirement which is not always easy to obtain in realistic theories and makes the inflationary model building problem a difficult issue [39]. The physical nature of the inflaton field has not been identified (there are many candidates) and, as a consequence, the shape of $V(\varphi)$ is not known. Of course, different $V(\varphi)$ lead to different inflationary expansions but, since these different potentials must all be sufficiently flat, the corresponding scale factors are all approximately given by the de Sitter solution. This solution is described by the scale factor $a(t) \simeq e^{Ht}$, where $H \equiv \dot{a}/a$ is the Hubble parameter, a slowly-varying quantity directly related to the energy scale of inflation. Observationally, this latter quantity is not known but is constrained [22] to be between the Grand Unified Theory (GUT) scale, that is to say $\sim 10^{15}$ GeV, and ~ 1 TeV. It is also interesting to remark that the energy density during inflation, $\rho_{\rm inf} \simeq 3H_{\rm inf}^2 M_{\rm Pl}^2$, remains approximately constant. This implies the remarkable property that the classical background approximation remains valid even if one runs the clock backwards and study the initial stages of inflation. This would no longer be the case if an ordinary fluid (*e.g.* radiation or matter) dominated the energy budget of the early Universe since the scaling $\rho \propto 1/a^4$ or $\rho \propto 1/a^3$ would quickly imply that $\rho \gg M_{\rm Pl}^4$ at early times. Of course, this does not mean that quantum effects play no role during inflation. As a matter of fact, when the quantum kicks $\Delta\varphi_{\rm q}$ are much larger than the typical classical motion $\Delta\varphi_{\rm cl} \simeq V'/(3H^2)$, they must be taken into account. This can be done in the framework of stochastic inflation [40, 41].

But the arguably most impressive result coming from including the quantum effects into the inflationary picture concerns the origin of the cosmological fluctuations that give rise to the Cosmic Microwave Background (CMB) anisotropies and to the large-scale structures. We turn to this question in the next section.

3. Inflationary Perturbations

According to inflation, the seeds of cosmological fluctuations are the quantum fluctuations of the inflaton and gravitational fields. As is well-known, this idea leads to predictions that are in remarkable agreement with astrophysical data [20, 21, 22, 23, 24, 25, 26]. From the theoretical point of view, the perturbations are viewed as small ripples on top of a classical background in close analogy with the condensed matter situation where phonons are considered to be quantized excitations propagating within a classical crystal. In order to describe cosmological perturbations, one has to go beyond the homogeneity and isotropy of the FRLW metric. This can be accomplished by writing the metric as [12] $\mathrm{d}s^2 = a^2(\eta)\Big\{-(1-2\phi)\,\mathrm{d}\eta^2 + 2(\partial_i B)\,\mathrm{d}x^i\mathrm{d}\eta + [(1-2\psi)\,\delta_{ij} + 2\partial_i\partial_j E]\,\mathrm{d}x^i\mathrm{d}x^j\Big\}$, where η is the conformal time and where the four functions ϕ, B, ψ and E are functions of time and space. The above approach is, however, redundant because of gauge freedom [12, 42, 43]. The gravitational sector can in fact be described by a single, gauge-invariant, quantity, the so-called Bardeen potential Φ_{B} defined by [42] $\Phi_{\mathrm{B}}(\eta, \boldsymbol{x}) = \phi + [a(B-E')]'/a$, where a prime denotes a derivative with respect to the conformal time η. In the same manner, the matter sector can be modeled by the gauge invariant fluctuation of the scalar field $\delta\varphi^{(\mathrm{gi})}(\eta, \boldsymbol{x}) = \delta\varphi + \varphi'(B-E')$. The two quantities Φ_{B} and $\delta\varphi^{(\mathrm{gi})}$ are related by a perturbed Einstein constraint. This implies that the scalar sector can in fact be described by a single quantity. This quantity is the Mukhanov-Sasaki variable [6, 44] which is a combination of the Bardeen potential and of the gauge invariant field, namely $v(\eta, \boldsymbol{x}) = a\left[\delta\varphi^{(\mathrm{gi})} + \varphi'\Phi_{\mathrm{B}}/\mathcal{H}\right]$, where $\mathcal{H} \equiv a'/a$. It is also convenient to Fourier transform $v(\eta, \boldsymbol{x})$ since we consider a linear theory and, hence, all the modes evolve independently. Therefore, instead of $v(\eta, \boldsymbol{x})$, we will rather work with its Fourier amplitude $v_{\boldsymbol{k}}(\eta)$. Its equation of motion can be easily obtained from the perturbed Einstein equations and reads

$$v_{\boldsymbol{k}}'' + \omega^2(\eta, \boldsymbol{k})\, v_{\boldsymbol{k}} = 0, \tag{1}$$

where the time-dependent frequency $\omega(\eta, \boldsymbol{k})$ can be written as $\omega^2(\eta, \boldsymbol{k}) = k^2 - \left(a\sqrt{\epsilon_1}\right)''/(a\sqrt{\epsilon_1})$, k being the wavenumber of the mode under consideration and $\epsilon_1 \equiv -\dot{H}/H^2$ the first slow-roll parameter characterizing the cosmological expansion during inflation. The above equation is the equation of a parametric oscillator (*i.e.* similar to a pendulum the length of which can change with time). In the Minkowski situation, the scale factor is a constant and $\omega = k$. In this case, one deals with an harmonic oscillator. We see that the time dependence of the frequency $\omega(\eta, \boldsymbol{k})$ is entirely due to the fact that the perturbations live in a dynamical spacetime. It is worth noticing that Eq. (1) appears in other interesting physical situations such as the Schwinger effect [45, 46] or the dynamical Casimir effect [47] (recently observed for the first time in the laboratory, see Ref. [48]) and, each time, the same phenomenon is observed: particle creation under the influence of the classical, time-dependent, source even if the details of the corresponding process (number of pairs created, power spectrum etc ...) depend on the precise time evolution of the source. In this sense, the theory of cosmological perturbations can be viewed as "standard" in the framework of quantum field theory. The only new aspect is that the time-dependent source is the classical gravitational background of the expanding space-time. Let us also notice that inflation is in fact not necessary in order to have particle creation: as is clear from the above considerations, only a time-dependent scale factor is necessary. However, the inflationary expansion becomes mandatory if one wants a $\omega(\eta, \boldsymbol{k})$ which leads to a power spectrum in agreement with the CMB measurements.

We now describe how the cosmological fluctuations are quantized. In order to achieve this goal, it is also more convenient to work with real variables. Therefore, we introduce the following definitions $v_{\boldsymbol{k}} \equiv \left(v_{\boldsymbol{k}}^{\mathrm{R}} + i v_{\boldsymbol{k}}^{\mathrm{I}}\right)/\sqrt{2}$ and for the conjugate momentum of the Mukhanov-Sasaki variable, $p_{\boldsymbol{k}} \equiv \left(p_{\boldsymbol{k}}^{\mathrm{R}} + i p_{\boldsymbol{k}}^{\mathrm{I}}\right)/\sqrt{2}$. In the Schrödinger approach, the quantum state of the system is described by a wavefunctional, $\Psi\left[v(\eta, \boldsymbol{x})\right]$. Since we are working in Fourier space (and since the theory is still free in the sense that it does not contain terms with power higher than two in the Lagrangian), the wavefunctional can also be factorized into mode components as

$$\Psi\left[v(\eta, \boldsymbol{x})\right] = \prod_{\boldsymbol{k}} \Psi_{\boldsymbol{k}}\left(v_{\boldsymbol{k}}^{\mathrm{R}}, v_{\boldsymbol{k}}^{\mathrm{I}}\right) = \prod_{\boldsymbol{k}} \Psi_{\boldsymbol{k}}^{\mathrm{R}}\left(v_{\boldsymbol{k}}^{\mathrm{R}}\right) \Psi_{\boldsymbol{k}}^{\mathrm{I}}\left(v_{\boldsymbol{k}}^{\mathrm{I}}\right). \tag{2}$$

It can be shown that the solution to the Schrödinger equation can be written as a Gaussian function, which should not come as a surprise since we are quantizing an oscillator. Concretely,

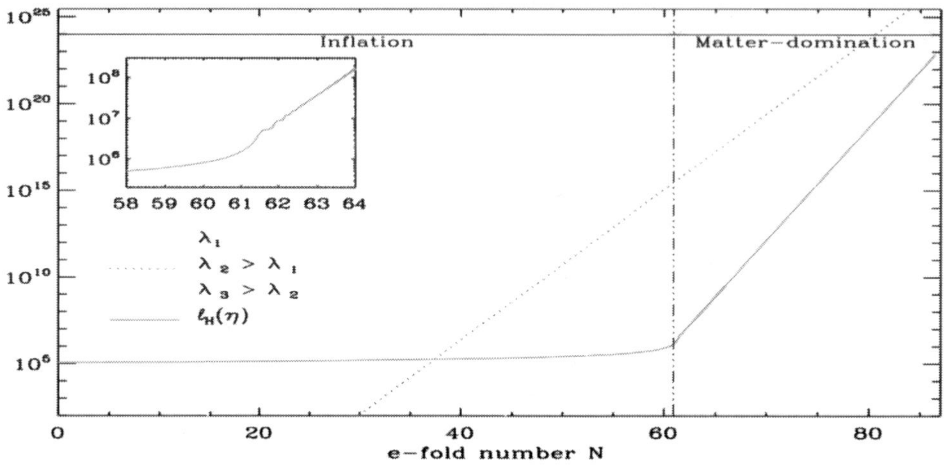

Figure 1. Evolution of various relevant length scales (in Planck units) during and after inflation. The solid red line represents the Hubble radius $\ell_{\rm H} \equiv 1/H$. It is almost constant during inflation and starts to grow during the post-inflation era (during this phase, the inflaton field oscillates at the bottom of its potential and the universe becomes matter-dominated. Small oscillations in $\ell_{\rm H}$ are clearly visible in the inset). The three dashed lines represent the physical wavelengths of three different Fourier modes. Initially within the Hubble radius, they exit $\ell_{\rm H}$ during inflation and re-enter it in the post-inflationary era.

one can write

$$\Psi_{\boldsymbol{k}}^{\rm R,I}\left(\eta, v_{\boldsymbol{k}}^{\rm R,I}\right) = N_{\boldsymbol{k}}(\eta) e^{-\Omega_{\boldsymbol{k}}(\eta)\left(v_{\boldsymbol{k}}^{\rm R,I}\right)^2}. \tag{3}$$

where the functions $N_{\boldsymbol{k}}(\eta)$ and $\Omega_{\boldsymbol{k}}(\eta)$ (which do not depend on whether one considers the wavefunction of the real or imaginary parts of the Mukhanov-Sasaki variable) can be expressed as

$$|N_{\boldsymbol{k}}| = \left(\frac{2\Re\mathrm{e}\,\Omega_{\boldsymbol{k}}}{\pi}\right)^{1/4}, \quad \Omega_{\boldsymbol{k}} = -\frac{i}{2}\frac{f_{\boldsymbol{k}}'}{f_{\boldsymbol{k}}}, \tag{4}$$

where $f_{\boldsymbol{k}}$ is a function obeying the equation $f_{\boldsymbol{k}}'' + \omega^2 f_{\boldsymbol{k}} = 0$, that is to say exactly Eq. (1).

We now discuss the question of the initial conditions. The fundamental assumption of inflation is that the perturbations are initially in their ground state. At the beginning of inflation, all the modes of astrophysical interest today have a physical wavelength smaller than the Hubble radius, see Fig. 1. In this regime, each mode now behaves as a harmonic oscillator (as opposed to a parametric oscillator in the generic case) with frequency $\omega = k$. As a consequence, the differential equation for $f_{\boldsymbol{k}}(\eta)$ can easily be solved and the solution reads $f_{\boldsymbol{k}} = A_{\boldsymbol{k}} e^{ik\eta} + B_{\boldsymbol{k}} e^{-ik\eta}$, $A_{\boldsymbol{k}}$ and $B_{\boldsymbol{k}}$ being integration constants. Upon using the second equation (4), one has

$$\Omega_{\boldsymbol{k}} \to \frac{k}{2}\frac{A_{\boldsymbol{k}} e^{ik\eta} - B_{\boldsymbol{k}} e^{-ik\eta}}{A_{\boldsymbol{k}} e^{ik\eta} + B_{\boldsymbol{k}} e^{-ik\eta}}. \tag{5}$$

The wavefunction (3) represents the ground state wavefunction of a harmonic oscillator if $\Omega_{\boldsymbol{k}} = k/2$. Therefore, one must choose the initial conditions such that $B_{\boldsymbol{k}} = 0$, the value of $A_{\boldsymbol{k}}$ turning out to be irrelevant in this regime. This choice of initial conditions completely determines the solution and is referred to as the Bunch-Davies vacuum [49].

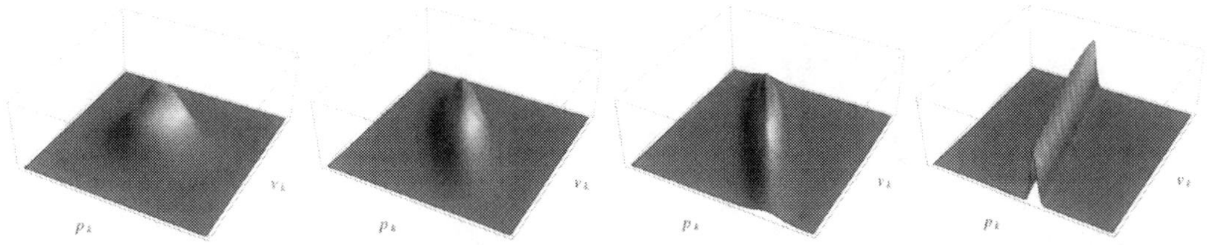

Figure 2. Wigner function of a squeezed quantum state at different times during inflation. Only the two-dimensional function corresponding to the set of variables $(v_{\boldsymbol{k}}^{\rm R}, p_{\boldsymbol{k}}^{\rm R})$ has been represented, see Eq. (7). The time evolution of $\Re{\rm e}\,\Omega_{\boldsymbol{k}}$ and $\Im{\rm m}\,\Omega_{\boldsymbol{k}}$ has been calculated using the Schrödinger equation with the Hamiltonian that can be obtained from the relativistic theory of cosmological perturbations. The left panel corresponds to $r_{\boldsymbol{k}} = 0.0005$ and the corresponding state is almost a coherent one. Then, from left to right, one has $r_{\boldsymbol{k}} = 0.48$, $r_{\boldsymbol{k}} = 0.88$ and, finally, $r_{\boldsymbol{k}} = 2.31$ for the right panel. The effect of the squeezing and the cigar shape of Eq. (7) are clearly visible.

We have just seen that, when the wavelength of a Fourier mode is smaller than the Hubble radius, the corresponding wavefunction can be chosen to be the ground state wavefunction of a harmonic oscillator. However, this is no longer the case when the mode exits the Hubble radius, see Fig. 1. In this regime, the function $\Omega_{\boldsymbol{k}}(\eta)$ no longer equals $k/2$ and acquires a non trivial time-dependence. In fact, a time dependent wavefunction as the one in Eq. (3) represents a very peculiar quantum state known as a squeezed state. A squeezed state is a Gaussian state for which there exists a direction in the plane $(p_{\boldsymbol{k}}, v_{\boldsymbol{k}})$ where the dispersion is extremely small, *i.e.* exponentially small $\simeq {\rm e}^{-2r_{\boldsymbol{k}}}$ where $r_{\boldsymbol{k}}$ is the squeezing parameter (of course, the dispersion in the perpendicular direction is very large $\simeq {\rm e}^{2r_{\boldsymbol{k}}}$ in order to satisfy the Heisenberg relation). The phenomenon of squeezing is widely studied in many different branches of physics, in particular in quantum optics. Squeezing occurs each time the quantization of a parametric oscillator is carried out. It is remarkable that the quantization of small fluctuations on top of an expanding universe also leads to that phenomenon. Another interesting feature is that cosmological squeezing is much larger than what can be achieved in the laboratory [50]: for modes of cosmological interest today, $r_{\boldsymbol{k}}$ can reach values of order $\sim 10^2$ [27]. In the literature, this regime is very often described as a regime where the cosmological perturbations have classicalized [30, 29, 33, 33, 51]. Since this concept is subtle in quantum mechanics, we need to come back to this issue and to describe accurately what is meant by a "classical limit" in this context. In particular, it may seem strange at first sight that a quantum system placed in a strongly squeezed state can be described as a classical state since, in the context of, say, quantum optics, a similar situation would precisely be described as a non-classical situation [52, 53].

A convenient tool to study this question is the Wigner function, defined by

$$W\left(v_{\boldsymbol{k}}^{\rm R}, v_{\boldsymbol{k}}^{\rm I}, p_{\boldsymbol{k}}^{\rm R}, p_{\boldsymbol{k}}^{\rm I}\right) = \frac{1}{(2\pi)^2}\int {\rm d}x{\rm d}y\, \Psi^*\left(v_{\boldsymbol{k}}^{\rm R} - \frac{x}{2}, v_{\boldsymbol{k}}^{\rm I} - \frac{y}{2}\right) {\rm e}^{-ip_{\boldsymbol{k}}^{\rm R}x - ip_{\boldsymbol{k}}^{\rm I}y}\, \Psi\left(v_{\boldsymbol{k}}^{\rm R} + \frac{x}{2}, v_{\boldsymbol{k}}^{\rm I} + \frac{y}{2}\right). \quad (6)$$

It is well known that the Wigner function can be understood as a classical probability distribution function whenever it is positive definite. Then, upon using the quantum state (3), the following explicit form is obtained

$$W\left(v_{\boldsymbol{k}}^{\rm R}, v_{\boldsymbol{k}}^{\rm I}, p_{\boldsymbol{k}}^{\rm R}, p_{\boldsymbol{k}}^{\rm I}\right) = \Psi\Psi^* \frac{1}{2\pi \Re{\rm e}\,\Omega_{\boldsymbol{k}}} \exp\left[-\frac{1}{2\Re{\rm e}\,\Omega_{\boldsymbol{k}}}\left(p_{\boldsymbol{k}}^{\rm R} + 2\Im{\rm m}\,\Omega_{\boldsymbol{k}} v_{\boldsymbol{k}}^{\rm R}\right)^2\right]$$
$$\times \exp\left[-\frac{1}{2\Re{\rm e}\,\Omega_{\boldsymbol{k}}}\left(p_{\boldsymbol{k}}^{\rm I} + 2\Im{\rm m}\,\Omega_{\boldsymbol{k}} v_{\boldsymbol{k}}^{\rm I}\right)^2\right]. \quad (7)$$

This Wigner function (7) is represented in Fig. 2 at different times or, equivalently, at different values of $r_{\boldsymbol{k}}$ ($r_{\boldsymbol{k}} = 0.0005$, 0.48, 0.88 and 2.31). The effect of strong squeezing is clearly visible. Initially, in the sub-Hubble regime, $r_{\boldsymbol{k}}$ is small and the Wigner function is peaked over of small region in phase space. In this regime, W is just the Wigner function of a coherent state since we chose to start from the ground state of a harmonic oscillator. Coherent states are considered as the "most classical" states precisely for this reason: if one is given, say, the value of $v_{\boldsymbol{k}}^{\mathrm{R}}$, then one obtains a value for the momentum, $p_{\boldsymbol{k}}^{\mathrm{R}}$, which is very close to the one one would have inferred in the classical case. This is of course due to the fact that the Wigner function follows the classical trajectory and has minimal spread around it in all phase space directions.

As inflation proceeds, the situation described before changes. The modes become super Hubble and $r_{\boldsymbol{k}}(\eta)$ increases. As a consequence, the Wigner function spreads and acquires a cigar shape typical of squeezed states [27]. Therefore, if one is now given $p_{\boldsymbol{k}}^{\mathrm{R}}$ then the value of $v_{\boldsymbol{k}}^{\mathrm{R}}$ is very uncertain since the Wigner function is spread over a large region in phase space. In this case, cosmological perturbations certainly do not behave classically in the usual sense. Given the previous discussion, it may seem relatively easy to observe genuine quantum effects in the CMB. Unfortunately this is not so, essentially because, in the strong squeezed limit, all quantum predictions can be in fact obtained from averages performed by mean of a classical stochastic process. The point is that, in the limit $r_{\boldsymbol{k}} \to \infty$, all the quantum predictions can be reproduced if one assumes that the system always followed classical laws but had random initial conditions with a given probability density function. This is possible because the Wigner function, despite being delocalized over phase space, remains positive. This property is not general but is a key feature of Gaussian states, hence of squeezed states.

Let us say a few words about the density matrix $\hat{\rho}_{\boldsymbol{k}}^{\mathrm{R}}$. The density matrix is nothing but the Fourier transform of the Wigner function. Let us denote by $|v_{\boldsymbol{k}}^{\mathrm{R}}\rangle$ the eigenstates of the operator $\hat{v}_{\boldsymbol{k}}^{\mathrm{R}}$. Then, we have

$$\langle v_{\boldsymbol{k}}^{\mathrm{R}\prime}|\hat{\rho}_{\boldsymbol{k}}^{\mathrm{R}}|v_{\boldsymbol{k}}^{\mathrm{R}}\rangle = \int_{-\infty}^{\infty} \mathrm{d}y\, \mathrm{e}^{iy(v_{\boldsymbol{k}}^{\mathrm{R}\prime}-v_{\boldsymbol{k}}^{\mathrm{R}})} W\left(\frac{v_{\boldsymbol{k}}^{\mathrm{R}\prime}+v_{\boldsymbol{k}}^{\mathrm{R}}}{2}, y\right). \tag{8}$$

Upon using Eq. (7) in the above equation, one obtains the following expression

$$\langle v_{\boldsymbol{k}}^{\mathrm{R}\prime}|\hat{\rho}_{\boldsymbol{k}}^{\mathrm{R}}|v_{\boldsymbol{k}}^{\mathrm{R}}\rangle = \left(\frac{2\Re\mathrm{e}\,\Omega_{\boldsymbol{k}}}{\pi}\right)^{1/2} \mathrm{e}^{-\Re\mathrm{e}\,\Omega_{\boldsymbol{k}}\left[(v_{\boldsymbol{k}}^{\mathrm{R}\prime})^2+(v_{\boldsymbol{k}}^{\mathrm{R}})^2\right]} \mathrm{e}^{-i\Im\mathrm{m}\,\Omega_{\boldsymbol{k}}\left[(v_{\boldsymbol{k}}^{\mathrm{R}\prime})^2-(v_{\boldsymbol{k}}^{\mathrm{R}})^2\right]}. \tag{9}$$

We notice that the off-diagonal terms, $v_{\boldsymbol{k}}^{\mathrm{R}\prime} \neq v_{\boldsymbol{k}}^{\mathrm{R}}$, oscillate very rapidly in the strong squeezing limit where $\Im\mathrm{m}\,\Omega_{\boldsymbol{k}} \to \infty$. This means that decoherence (defined as the disappearance of those off-diagonal terms) does not occur without taking into account the influence of an environment for the perturbations. Various discussions on what this environment may be can be found in Refs. [35, 36, 37].

We end this section by a brief discussion of how the theory presented above can make astrophysical predictions. The perturbations described by the Mukhanov-Sasaki variable are directly related to CMB fluctuations. As a consequence, in order to calculate the statistical properties of the CMB, it is sufficient to evaluate the power spectrum of the operator $\hat{v}(\eta, \boldsymbol{x})$ (and the higher order correlation functions if one is interested in non-Gaussianity). It is given by the following expression

$$\langle\Psi|\hat{v}(\eta,\boldsymbol{x})\hat{v}(\eta,\boldsymbol{x}+\boldsymbol{r})|\Psi\rangle = \int\prod_{\boldsymbol{k}} \mathrm{d}v_{\boldsymbol{k}}^{\mathrm{R}}\mathrm{d}v_{\boldsymbol{k}}^{\mathrm{I}}\Psi_{\boldsymbol{k}}^*(v_{\boldsymbol{k}}^{\mathrm{R}}, v_{\boldsymbol{k}}^{\mathrm{I}})v(\eta,\boldsymbol{x})v(\eta,\boldsymbol{x}+\boldsymbol{r})\Psi_{\boldsymbol{k}}(v_{\boldsymbol{k}}^{\mathrm{R}}, v_{\boldsymbol{k}}^{\mathrm{I}}). \tag{10}$$

and, upon using the results exposed before, can be expressed as

$$\langle\Psi|\hat{v}(\eta,\boldsymbol{x})\hat{v}(\eta,\boldsymbol{x}+\boldsymbol{r})|\Psi\rangle = \int_0^{+\infty}\frac{\mathrm{d}k}{k}\frac{\sin kr}{kr}\mathcal{P}_v(k), \tag{11}$$

which, in fact, defines the power spectrum $\mathcal{P}_v(k)$ as the square of the Fourier amplitude per logarithmic interval at a given scale, *i.e.* $\mathcal{P}_v(k) = k^3|f_{\boldsymbol{k}}|^2/(2\pi^2)$. On large scales, one obtains $\mathcal{P}_\zeta(k) \simeq A_{\rm S} k^{n_{\rm S}-1}$ where the coefficients $A_{\rm S}$ and $n_{\rm S}$ (the spectral index) depend on the inflationary model under consideration. For any "good" model of inflation, one has $n_{\rm S} \simeq 1$ but $n_{\rm S} \neq 1$. This means that inflation does not predict a scale-invariant power spectrum as sometimes said but an almost scale-invariant power spectrum. This difference is not a detail as a scale-invariant power spectrum (or a Harrison-Zeldovitch power spectrum) was considered well before the advent of inflation. In this case, $n_{\rm S} = 1$ would just be a postdiction. On the contrary, $n_{\rm S} \simeq 1$ but $n_{\rm S} \neq 1$ is a prediction of inflation. This prediction has not yet be confirmed by the observations but there is good evidence that it is true. For instance, the WMAP data indicate that $1 - n_{\rm S} = 0.018^{+0.019}_{-0.02}$ [20, 21, 22]). In a few months, with the data of the Planck satellite, this prediction will be definitively confirmed or ruled out with certainty (*i.e.* at the 5σ confidence level). Of course, the spectral index is not the only quantity of interest and a list and a discussion of all the inflationary predictions can be found in Refs. [13, 14, 15].

4. Conclusions

In this short letter, we have discussed the quantum-mechanical aspect of inflationary fluctuations. In particular, we have argued that cosmological fluctuations are placed in a very peculiar quantum state, namely a strongly squeezed state. In the very same way that the CMB radiation represents the most accurate black body radiation ever produced (recall that, in the laboratory, it is not possible to produced such an accurate black body), the squeezed state of the inflationary perturbations probably represents the strongest squeezed state ever observed in Nature. As a consequence, one could be led to conclude that the quantum properties of the CMB anisotropies are highly non-classical. And, in a sense, they are! But, we have also shown that everything turns out to be equivalent to a classical stochastic process. Therefore, even if the statistical nature of the fluctuations is here to remind us about their quantum origin, all the genuine quantum-mechanical imprints (such as interferences or the off-diagonal terms in the density matrix) are "washed out" by the strong squeezing and by decoherence. This explains why astronomers can forget about the quantum origin of the CMB fluctuations and consider that they are simply described by a Gaussian stochastic process.

There remains, however, an open question, namely the issue of definite outcomes. This issue is of course already present in conventional quantum mechanics (decoherence does not solve the definite outcome problem, see Ref. [54]) but it is clearly more embarrassing in the context of cosmology where no observer was present during inflation [38]. This problem was recently discussed in Ref. [55]. This is usually taken as an indication that the Copenhagen interpretation is not well suited to cosmology and that other theoretical frameworks, such as (for instance) the many worlds interpretation or the collapse theories, should be considered. This also shows that inflation is not only a scenario which allows us to convincingly reproduce various astrophysical observations but is also a playground where very fundamental questions can be discussed and studied. Usually this is considered to be a feature of quantum cosmology only but, in fact, this is also the case for inflation. It is clear that quantum effects in the context of inflation are treated in a perfectly standard way and, moreover, that we have at our disposal detailed observations that directly originate from those quantum effects. For this reason, it does not seem necessary to go as far as quantum cosmology (although it is of course interesting) to discuss very fundamental questions at the crossroads of quantum mechanics and gravity: inflation can do the job!

Acknowledgments

I would like to thank the organizers of the COSGRAV12 conference for their hospitality. I thank Marc Lilley and Vincent Vennin for careful reading of the manuscript.

References

[1] Starobinsky A A 1980 Phys. Lett. **B91** 99–102
[2] Guth A H 1981 Phys. Rev. **D23** 347–356
[3] Linde A D 1982 Phys. Lett. **B108** 389–393
[4] Albrecht A and Steinhardt P J 1982 Phys. Rev. Lett. **48** 1220–1223
[5] Linde A D 1983 Phys. Lett. **B129** 177–181
[6] Mukhanov V F and Chibisov G 1981 JETP Lett. **33** 532–535
[7] Mukhanov V F and Chibisov G 1982 Sov. Phys. JETP **56** 258–265
[8] Starobinsky A A 1982 Phys. Lett. **B117** 175–178
[9] Guth A H and Pi S Y 1982 Phys. Rev. Lett. **49** 1110–1113
[10] Hawking S 1982 Phys. Lett. **B115** 295 revised version
[11] Bardeen J M, Steinhardt P J and Turner M S 1983 Phys. Rev. **D28** 679
[12] Mukhanov V F, Feldman H and Brandenberger R H 1992 Phys.Rept. **215** 203–333
[13] Martin J 2004 Braz. J. Phys. **34** 1307–1321 (*Preprint* astro-ph/0312492)
[14] Martin J 2005 Lect. Notes Phys. **669** 199–244 (*Preprint* hep-th/0406011)
[15] Martin J 2008 Lect. Notes Phys. **738** 193–241 (*Preprint* 0704.3540)
[16] Linde A D 2008 Lect. Notes Phys. **738** 1–54 (*Preprint* 0705.0164)
[17] Sriramkumar L 2009 (*Preprint* 0904.4584)
[18] Peter P and Uzan J P 2009 Primordial Cosmology Oxford Graduate Texts (Oxford University Press)
[19] Stewart E D and Lyth D H 1993 Phys.Lett. **B302** 171–175 (*Preprint* gr-qc/9302019)
[20] Larson D, Dunkley J, Hinshaw G, Komatsu E, Nolta M et al. 2011 Astrophys.J.Suppl. **192** 16 (*Preprint* 1001.4635)
[21] Komatsu E et al. (WMAP Collaboration) 2011 Astrophys.J.Suppl. **192** 18 (*Preprint* 1001.4538)
[22] Martin J and Ringeval C 2006 JCAP **0608** 009 (*Preprint* astro-ph/0605367)
[23] Lorenz L, Martin J and Ringeval C 2008 JCAP **0804** 001 (*Preprint* 0709.3758)
[24] Lorenz L, Martin J and Ringeval C 2008 Phys.Rev. **D78** 063543 (*Preprint* 0807.2414)
[25] Martin J and Ringeval C 2010 Phys.Rev. **D82** 023511 (*Preprint* 1004.5525)
[26] Martin J, Ringeval C and Trotta R 2011 Phys.Rev. **D83** 063524 (*Preprint* 1009.4157)
[27] Grishchuk L and Sidorov Y 1990 Phys.Rev. **D42** 3413–3421
[28] Grishchuk L, Haus H and Bergman K 1992 Phys.Rev. **D46** 1440–1449
[29] Polarski D and Starobinsky A A 1996 Class.Quant.Grav. **13** 377–392 (*Preprint* gr-qc/9504030)
[30] Lesgourgues J, Polarski D and Starobinsky A A 1997 Nucl.Phys. **B497** 479–510 (*Preprint* gr-qc/9611019)
[31] Grishchuk L and Martin J 1997 Phys.Rev. **D56** 1924–1938 (*Preprint* gr-qc/9702018)
[32] Kiefer C, Lohmar I, Polarski D and Starobinsky A A 2007 Class.Quant.Grav. **24** 1699–1718 (*Preprint* astro-ph/0610700)
[33] Kiefer C, Polarski D and Starobinsky A A 1998 Int.J.Mod.Phys. **D7** 455–462 (*Preprint* gr-qc/9802003)
[34] Egusquiza I, Feinstein A, Perez Sebastian M and Valle Basagoiti M 1998 Class.Quant.Grav. **15** 1927–1936 (*Preprint* gr-qc/9709061)
[35] Anderson P R, Molina-Paris C and Mottola E 2005 Phys.Rev. **D72** 043515 (*Preprint* hep-th/0504134)
[36] Burgess C P, Holman R and Hoover D 2008 Phys.Rev. **D77** 063534 (*Preprint* astro-ph/0601646)
[37] Martineau P 2007 Class.Quant.Grav. **24** 5817–5834 (*Preprint* astro-ph/0601134)
[38] Sudarsky D 2011 Int.J.Mod.Phys. **D20** 509–552 (*Preprint* 0906.0315)
[39] Lyth D H and Riotto A 1999 Phys.Rept. **314** 1–146 (*Preprint* hep-ph/9807278)
[40] Starobinsky A A 1986
[41] Martin J and Musso M 2006 Phys. Rev. **D73** 043516 (*Preprint* hep-th/0511214)
[42] Bardeen J M 1980 Phys.Rev. **D22** 1882–1905
[43] Martin J and Schwarz D J 1998 Phys.Rev. **D57** 3302–3316 (*Preprint* gr-qc/9704049)
[44] Kodama H and Sasaki M 1984 Prog.Theor.Phys.Suppl. **78** 1–166
[45] Schwinger J S 1951 Phys.Rev. **82** 664–679
[46] Brezin E and Itzykson C 1970 Phys.Rev. **D2** 1191–1199
[47] Moore G T 1970 Journal of Mathematical Physics **11** 2679–2691
[48] Wilson C M, Johansson G, Pourkabirian A, Simoen M, Johansson J R, Duty T, Nori F and Delsing P 2011 Nature **479** 376–379 (*Preprint* 1105.4714)
[49] Birrell N D and Davies P C W 1982 Quantum Fields In Curved Space (Cambridge Univ. Pr.)
[50] Vahlbruch H, Mehmet M, Chelkowski S, Hage B, Franzen A, Lastzka N, Goßler S, Danzmann K and Schnabel R 2008 Physical Review Letters **100** 033602 (*Preprint* 0706.1431)
[51] Kiefer C 2000 Nucl.Phys.Proc.Suppl. **88** 255–258 (*Preprint* astro-ph/0006252)
[52] Caves C M and Schumaker B L 1985 Phys.Rev. **A31** 3068–3092
[53] Schumaker B L and Caves C M 1985 Phys.Rev. **A31** 3093–3111

[54] Colanero K 2012 ArXiv e-prints (*Preprint* 1208.0904)
[55] Martin J, Vennin V and Peter P 2012 (*Preprint* 1207.2086)

Black holes and holography

Samir D. Mathur

Department of Physics, The Ohio State University, 191 W. Woodruff Ave, Columbus, OH 43220, USA

E-mail: mathur.16@osu.edu

Abstract. The idea of holography in gravity arose from the fact that the entropy of black holes is given by their surface area. The holography encountered in gauge/gravity duality has no such relation however; the boundary surface can be placed at an arbitrary location in AdS space and its area does not give the entropy of the bulk. The essential issues are also different between the two cases: in black holes we get Hawking radiation from the 'holographic surface' which leads to the information issue, while in gauge/gravity duality there is no such radiation. To resolve the information paradox we need to show that there are real degrees of freedom at the horizon of the hole; this is achieved by the fuzzball construction. In gauge/gravity duality we have instead a field theory defined on an abstract dual space; there are no gravitational degrees of freedom at the holographic boundary. It is important to understand the relations and differences between these two notions of holography to get a full understanding of the lessons from the information paradox.

1. Introduction

The idea of holography arose in gravitational physics from the expression for the entropy of black holes [1]

$$S_{bek} = \frac{A}{4G} \tag{1}$$

Since the entropy is given by its surface area measured in units of l_p^2, it appeared plausible that there is 'one bit of data per planck area of the horizon'. Because the degrees of freedom are given by the bounding area rather than the 'volume' of the hole, we use the term 'holographic' to characterize the gravitational physics of the hole.

In recent years the term 'holography' has been applied to the idea of gauge/gravity duality in AdS spacetimes [2]. This idea is depicted in fig.1. The dashed line in fig.1(a) is an arbitrary boundary surface $\mathcal{S}$. A field theory on $\mathcal{S}$ describes the gravitational physics of the entire region below $\mathcal{S}$. Again, it appears that physics is 'holographic'.

Since the idea of gauge/gravity duality arose from studies of black holes in string theory, it is often assumed that these two uses of the term 'holographic' are the same. Stretching this further, one may think that the idea of gauge/gravity duality would somehow explain the mysteries associated with the entropy (1) and the related problem of the information paradox [3]. But as we now note, there are several differences between the above two notions of holography:

(i) For the black hole, the location of the boundary surface (the horizon) is fixed at $r = 2M$. In gauge/gravity duality, the boundary surface can be placed anywhere in the AdS region.

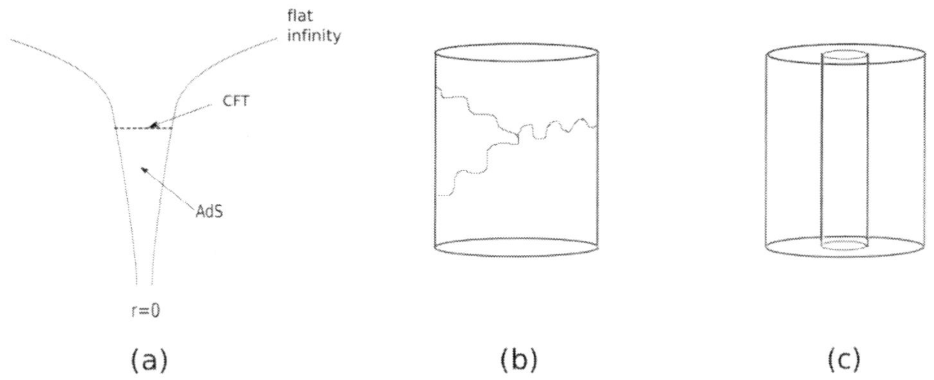

Figure 1. (a) Branes create a geometry that is AdS in the 'near' region; the dual CFT lives on a boundary placed anywhere in the AdS region, and describes gravity in all the region below it. (b) The singularity at $r = 0$ can be avoided by moving to global AdS, where a 3-point function is computed by a simple path integral with no singularities. (c) If we have enough energy in global AdS, we make a black hole, and then we face the difficulties of Hawking's argument again. (The vertical direction is time, and the surface of the inner cylinder is the black hole horizon.)

(ii) For the black hole, the area of the boundary surface gives the entropy (1) of the interior. But in gauge/gravity duality here is no such relation. The entropy of the interior spacetime depends on how much energy E we put in it. In particular if we take $E = 0$, we get empty AdS with entropy $S = 0$.

As we probe these differences deeper, we will uncover important aspects of gravitational degrees of freedom and of the structure of spacetime itself.

2. Lessons from the information paradox

A fundamental problem in black hole physics is the information paradox [3]. In a geometry with horizon, there is no time-independent foliation of spacetime. As shown in fig.2, spacelike slices are $t = constant$ outside the horizon, but $r = constant$ inside. 'Later' slices are obtained by 'stretching' the $r = constant$ part of the slice; this stretching creates particle pairs, with one member being inside the horizon and one outside. The important aspect of this particle creation process is that the two members of the pair are in an *entangled* state, which we may write schematically as

$$|\psi\rangle_{pair} = \frac{1}{\sqrt{2}}\left(|0\rangle_{in}|0\rangle_{out} + |1\rangle_{in}|1\rangle_{out}\right) \tag{2}$$

where $|0\rangle, |1\rangle$ represent occupation numbers 0, 1 respectively for a given particle mode. Thus each step of the pair creation process generates an entanglement entropy $S_{ent} = \ln 2$, and after N steps of particle creation the entanglement is

$$S_{ent} = N \ln 2 \tag{3}$$

This pair creation continues until the hole reaches planck size, at which point the emitted radiation is heavily entangled with the quanta in the hole. If the hole evaporates completely, the radiation cannot be attributed *any* quantum state – the radiation is entangled, but there is nothing that it is entangled *with*. This possibility is termed 'information loss' or 'loss of unitarity'. If we are left with a remnant, then this remnant needs to have at least 2^N internal states to permit the entanglement (3). Since N is unbounded, our theory must permit arbitrarily

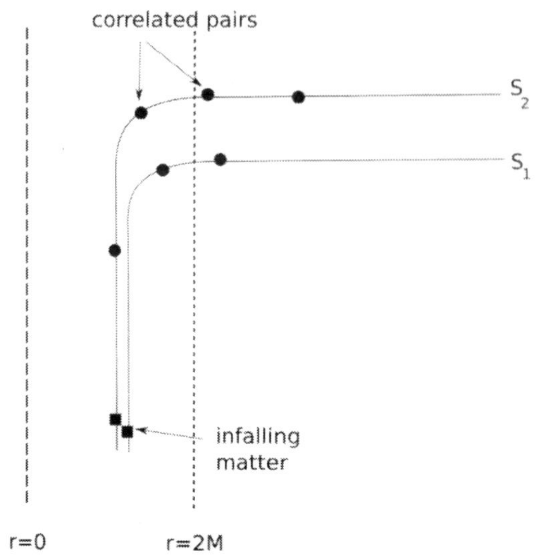

Figure 2. Eddington-Finkelstein coordinates for the Schwarzschild hole. Spacelike slices are $t = const$ outside the horizon and $r = const$ inside. Curvature length scale for a solar mass black hole is $\sim 3\ km$ all over the region of evolution covered by the slices S_i.

high degeneracy in a bounded volume with bounded energy budget; something that is hard to achieve with normal quantum theories.

Many relativists had reconciled themselves to admitting remnants of some sort, perhaps with baby Universes opening up inside the remnant where states could be hidden [4]. But such remnants have not been found in string theory. How do we avoid being forced to 'information loss', which would imply a breakdown of string theory?

Some string theorists have been seriously worried about this problem. But many others assumed that Hawking's argument was somehow *flawed*. Among the latter, the most common belief was the following. Hawking computed the pair creation at leading order, but there can always be small quantum gravity corrections to the wavefunction (2)

$$|\psi\rangle_{pair} = \frac{1}{\sqrt{2}}\left(|0\rangle_{in}|0\rangle_{out} + |1\rangle_{in}|1\rangle_{out}\right) + \epsilon\frac{1}{\sqrt{2}}\left(|0\rangle_{in}|0\rangle_{out} - |1\rangle_{in}|1\rangle_{out}\right) \qquad (4)$$

where we have added a small amount of an orthogonal state for the pair. The correction ϵ for each pair must be small since the horizon geometry is smooth, but the number of emitted quanta is large ($\sim (M/m_p)^2$), and the net effect of the small corrections may cumulate in such a way that the overall state of the radiation would be un-entangled with the hole.

But in [5] it was shown that this hope is false; the change in entanglement is bounded as

$$\frac{\delta S_{ent}}{S_{ent}} < 2\epsilon \qquad (5)$$

This inequality is the essential reason why the Hawking argument has proved so robust over the years – no small corrections can save the situation.

To summarize, it can be rigorously argued that we must choose between one of the following: (i) we have information loss or remnants (ii) we find a way to get order *unity* corrections to low energy physics at the horizon.

3. Gauge/gravity duality and the information paradox

With the above knowledge of the information problem, let us return to our analysis of holography. A common error is to argue the following: "Many computations support the idea that gravity is dual to a gauge theory. Since the latter is unitary, there cannot be information loss in black holes. Thus we have solved the information problem".

As we will now see, this argument is completely incorrect, and arises from ignoring the power of the information paradox, encoded in (5). We can approach gauge/gravity duality from two sides:

(i) We know string theory at low energy gives gravity, which contains black holes. Low energy gravity ampltiudes can be reproduced by the gauge theory. But if we put together enough energy to make an AdS-Schwarzschild black hole, then the Hawking argument tells us that we will get information loss. Thus unless we find some way to bypass the Hawking argument, *gauge/gravity duality would fail at the same place where all other quantum gravity approaches fail: at the threshold of black hole formation.*

(ii) We can *define* the gravity theory as the dual of the gauge theory. In this case we cannot lose unitarity. *But now we cannot assume that the dual gravity theory has black holes.* Low energy amplitudes in the gauge theory agree with gravity, but if we take a large energy excitation in the gauge theory then the natural timescale for dispersion of this energy is the *crossing time* of the black hole, not the much longer Hawking evaporation time. But if the energy disperses in order crossing time, we have no black hole in the theory.

As a concrete manifestation of the above points, we can look at the simplest manifestation of a gauge/gravity type correspondence: the 1-d matrix model which is dual to 1+1 dimensional gravity [6]. The low energy gravity theory is 1+1 dimensional dilaton gravity, in which we can make a black hole by infalling matter [7]. We find information loss or remnants, depending on how we complete the theory at the planck scale. The low energy gravity theory can be reproduced by a dual 'matrix model' which lives in 0+1 dimensions. This matrix model is unitary, but if we try to make a black hole we fail: the energy of a collapsing shell bounces off the origin and returns in a time of order crossing time. The differences from the low energy theory are caused by higher order quantum corrections, which are small at low energies, but grow large enough at the black hole threshold to prevent black hole formation altogether [8].

In short, gauge/gravity duality has no direct bearing on the information problem. As we noted in the introduction, the 'holography' in gauge/gravity duality is not the same as the 'holography' that arises for black holes from the entropy formula (1). The holography of gauge/gravity duality can be verified for low energy correlators, but above the threshold of black hole formation the duality cannot answer any quantum gravity questions until we understand the nature of gravitational states in this domain. The traditional approach of writing down the AdS-Schwarzschild metric for the black hole allows us to define thermal averages in the dual gauge theory, but cannot tell us anything about how the information paradox is to be resolved. For the latter question, we need to understand black hole microstates themselves.

4. Resolving the information paradox – fuzzballs

In string theory, we have to make black hole states from the objects present in the theory – strings and branes. It turns out that the size of brane bound states grows with the coupling and with the number of branes in the bound state – in such a way that the size of the state is always order horizon size [9]. Thus we do not get a traditional horizon with vacuum in its vicinity, as was assumed in the Hawking computation. The emission of low energy modes can be therefore modified by order unity, as required to solve the information paradox.

Further progress along these lines is obtained by taking specific states of the hole and constructing their gravity description. In each case we see that no horizon forms. It is interesting to see these constructions in the context of the no-hair conjectures that suggested that the traditional black hole geometry was *unique*. In string theory we have extra compact dimensions, and a set of sources (branes). These objects are all crucial to the structure of microstates. A compact circle 'pinches off' before reaching the horizon, creating a smooth end to the spacetime geometry. This pinch-off generates a set of Kaluza-Klein monopoles and antimonopoles just outside the place where the horizon would have formed, and fluxes corresponding to brane charges wrap the topological cycles produced by this monopole structure. The simplest black hole is the extremal 2-charge hole made from D1 and D5 branes. For this hole we find the following:

(i) The number of extremal bound states of the D1D5 brane system can be counted by abstract topological methods, and give a microscopic entropy $S_{micro} = 4\pi\sqrt{n_1 n_5}$ [10].

(ii) If we assume a spherically symmetric ansatz and a trivial factorization of the compact directions, then the low energy supergravity action gives an extremal black hole with horizon, with a Bekenstein-Wald entropy $S_{bek} = 4\pi\sqrt{n_1 n_5}$ [11].

(iii) The actual microstates of the D1D5 system can be constructed. It is found that they are not spherically symmetric and the compact directions are locally nontrivially fibered, though the net monopole charge of these fibrations vanishes. The solutions have no horizon and no singularity [12].

(iv) The phase space of these horizonless gravitational solutions can be quantized, and yields the entropy $S = 4\pi\sqrt{n_1 n_5}$ [13].

(v) Though there is no horizon for any microstate, the region where the typical microstates exhibits their nontrivial structure has a boundary whose area A satisfies $A/G \sim \sqrt{n_1 n_5} \sim S_{bek}$ [14].

Work on more complicated extremal holes [15] has yielded a similar picture [16], though all microstates have not been constructed yet. Some families of nonextremal microstates have been constructed as well [17], and again they have no horizon or singularity. But they do have *ergoregions*, and the rate of ergoregion emission [18] agrees exactly with the Hawking radiation rate expected for these microstates [19]. But this time the radiation process does not lead to information loss; the radiation is similar to that from a normal warm body.

5. The fate of a collapsing shell

Let us pause for a moment to see what the above discussion says about holography. We again find a difference between the case of the black hole and the case of gauge/gravity duality:

(a) For the black hole, we have found that microstates do not have traditional horizons where we would have found a featureless vacuum around the around the location of the horizon. Instead, the information of the microstate is encoded in the detailed structure of the microstate at the location where the horizon would have been. In short, there are real degrees of freedom at the surface which we use to get the notion of holography.

(ii) In the case of gauge/gravity duality, there are no degrees of freedom apparent at the location of the boundary used for holography. This fact is related to the observation that the holographc boundary in this case can be moved to an arbitrary location in the AdS region.

We now turn to addressing a common question with black holes: what happens to a shell that is collapsing to make a black hole?

Consider a shell of mass M that is collapsing through its horizon radius $R \sim GM$. In ordinary 3+1 dimensional gravity the wavefunction of the shell moves in the expected way to smaller r, creating the traditional black hole geometry. But in string theory the $e^{S_{bek}}$ states of the hole give alternate wavefunctions with the same quantum numbers as the shell. There is a small amplitude for the wavefunction of the shell to tunnel into one of these microstate wavefunctions. We may estimate this amplitude as $\mathcal{A} \sim e^{-S_{gravity}}$ where $S_{gravity} = \frac{1}{16\pi G}\int \mathcal{R}\sqrt{-g}d^4x$ and we use $\sim GM$ for all length scales. This gives

$$S_{gravity} \sim \frac{1}{G}\int \mathcal{R}\sqrt{-g}\,d^4x \sim \frac{1}{G}\frac{1}{(GM)^2}(GM)^4 \sim GM^2 \sim \left(\frac{M}{m_p}\right)^2 \tag{6}$$

Thus $\mathcal{A} \sim e^{-(M/m_p)^2}$ is indeed tiny. But we must now multiply by the *number* of states that we can tunnel to, and this is given by $\mathcal{N} \sim e^{S_{bek}}$ where

$$S_{bek} = \frac{A}{4G} \sim \frac{(GM)^2}{G} \sim \left(\frac{M}{m_p}\right)^2 \tag{7}$$

Thus the smallness of the tunneling amplitude is offset by the remarkably large degeneracy of states that the black hole has [20]. The wavefunction of the shell tunnels into these 'fuzzball' states in a time much shorter than the evaporation time of the hole [21], and then these fuzzballs states radiate energy much like any other normal body. In short, the semiclassical approximation that leads to the standard black hole geometry gets invalidated by the large measure of phase space over which the wavefunction of the shell can spread.

6. The failure of the 'good slicing' argument

We can rephrase the above discussion in the following way. The information problem arose due to the 'good slicing' of the black hole, in which we keep the slice smooth while stretching it more and more (fig.2). This stretching creates the entangled pairs that lead to Hawking's paradox. But with the above estimate of tunneling rates, we have a different situation. On an early time slice we can indeed arrange the wavefunction so that it describes a semiclassical spacetime containing a collapsing shell. But the wavefunction of a later slice has to be obtained by evolution (using the Hamiltonian constraint) from the wavefunction on the initial slice. If we had a simple theory of quantum gravity, like canonically quantized general relativity, then evolution of the earlier slice would indeed give the 'stretched' slice. But in string theory the situation is different. We have an enormous space of alternative solution to the gravity theory, where for example the compact directions can 'pinch off' to make monopole pairs. We can take the wavefunction on the initial slice to be peaked around the smooth gently curved manifold, but the evolution will force this initial wavefunction to spread over the space of fuzzball solutions. Thus instead of getting the 'stretched' slice, we get a linear combination of fuzzballs, which then radiate from their surface just like any other warm body. We depict this spreading of the wavefunction in fig.3.

We can also use this picture to address some related questions. Marolf has discussed the information problem in the following language. Suppose we assume that the gravity theory is holographically dual to a boundary field theory. Then the boundary theory contains all the information in the bulk, and we can connect the bulk state at early times to the bulk state at late times by just evolving unitarily in the boundary theory. Thus it seems impossible to have information loss in any theory with a gravity dual.

While this argument helps to frame the information problem sharply, note that we cannot use it to argue away the paradox itself. The reasons are the same as those we discussed in section 3. If we start with some simple theory of gravity, then we do not know that it will have a holographic dual at the energies where black holes will form. If we start with the field theory and *define* bulk gravity as its dual, then we do not know that we will get black holes.

But Marolf's argument *can* be used to sharpen the puzzles arising from the paradox, as was done recently in a paper by Heemskerk, Marolf and Polchinski [23]. These authors asked if the dual field theory captured the state of a 'Schrodinger's cat' that was behind the horizon.

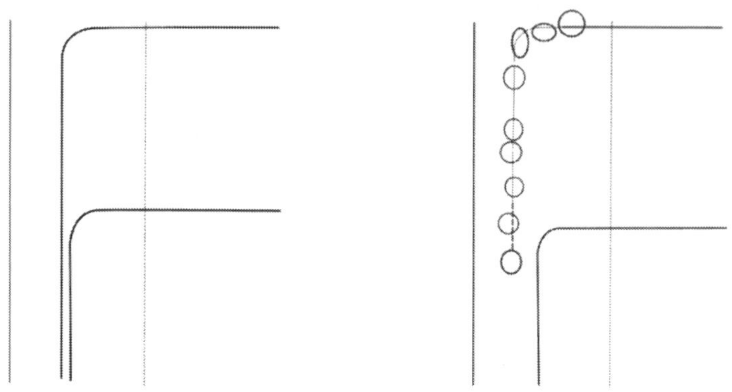

(a) (b)

Figure 3. (a) The stretching of 'good slices' in the traditional black hole geometry leads to pair creation by the Hawking process and the consequent information problem. (b) If there are $Exp[S]$ fuzzball solutions, the wavefunction giving semiclassical geometry on the initial slice spreads over this vast phase space of solutions after some evolution, and we no longer get the traditional pair creation with growing entanglement.

At early times when there was no black hole, the dual field theory presumably did capture all details of the bulk, and therefore captured the state of the cat. Evolution in the boundary is a known unitary evolution, so if the cat then evolved to be at a point inside the horizon, then its state would also be captured by the boundary theory. What appears puzzling though is that the radiation emitted from the hole should *also* capture the information of the cat, and one can draw a Cauchy slice that captures both this radiation and the cat behind the horizon. Since information cannot be duplicated, one expects a kind of 'complementarity' [24] which somehow does not create a contradiction between the presence of these two copies. This argument, of course, is just a restatement of the usual information problem of black holes; it appears sharper in the context of the dual field theory since this field theory is manifestly unitary and does not duplicate information.

With the help of fig.3 we can see how the puzzle is resolved with fuzzballs. Suppose the initial slice contains the collapsing shell and the cat, in their usual semiclassical wavefunctions. But the slice that captures a significant part of the radiation involves a lot of stretching, and this evolution spreads the wavefunction away from semiclassical slices to a linear combination of fuzzballs. The data of the cat is encoded in these fuzzballs, and is then carried away in the radiation from the fuzzballs, as in the process of ergoregion emission described above in section 4.

7. The infall problem and complementarity

It is interesting that we can get a notion of complementarity from the fuzzball picture. We have seen that spacetime 'ends' before the horizon is reached in black hole microstates. What happens when a heavy object ($E \gg kT$) falls onto the fuzzball? Thus question was addressed in [25], where it was argued that we excite collective modes of the fuzzball whose structure is relatively insensitive to the exact microstate that the hole was in. Further, applying ideas of Israel [26], Maldacena [27] and Van Raamsdonk [28], one can argue [25] that these collective modes will have the same spectrum as the spectrum of 'empty space with a horizon'. Thus low energy ($E \sim kT$) modes see the detailed structure of the microstate, allowing information to be carried out in Hawking radiation, while the approximate behavior of heavy infalling quanta can be approximated over the crossing timescale by the traditional geometry with vacuum around the horizon.

8. Summary

We have seen that the term 'holography' has been used in two different contexts in the study of gravity. In gauge/gravity duality, holography maps a gravity theory to an abstract theory at an arbitrarily placed boundary. There are no gravitational degrees of freedom at this boundary itself, its area does not give an entropy, and there is no radiation from this boundary leading to the Hawking problem. By contrast, the origins of holography lie in black hole thermodynamics, where the holographic boundary is the horizon. The horizon location is fixed, its area gives the entropy, and radiation from this horizon leads to the information paradox. If we confuse these two notions of holography, then we arrive at an erroneous conclusion that the information paradox is somehow magically evaded if we assume an imaginary surface at the horizon that carries the degrees of freedom of the hole. In actuality, the information paradox is evaded only because there are *real* degrees of freedom (hair) at the location of the horizon, and the construction of this hair is accomplished in string theory by the fuzzball construction.

Acknowledgements

This work was supported in part by DOE grant DE-FG02-91ER-40690.

References

[1] J. D. Bekenstein, Phys. Rev. D **7**, 2333 (1973).
[2] J. M. Maldacena, Adv. Theor. Math. Phys. **2**, 231 (1998) [Int. J. Theor. Phys. **38**, 1113 (1999)] [arXiv:hep-th/9711200]; E. Witten, Adv. Theor. Math. Phys. **2**, 253 (1998) [hep-th/9802150]; S. S. Gubser, I. R. Klebanov and A. M. Polyakov, Phys. Lett. B **428**, 105 (1998) [hep-th/9802109].
[3] S. W. Hawking, Commun. Math. Phys. **43**, 199 (1975) [Erratum-ibid. **46**, 206 (1976)]; S. W. Hawking, Phys. Rev. D **14**, 2460 (1976).
[4] J. Polchinski and A. Strominger, Phys. Rev. D **50** 7403 (1994), hep-th/9407008; T. Jacobson, gr-qc/0308048; D. A. Easson and R. H. Brandenberger, JHEP **0106**, 024 (2001), hep-th/0103019; A. Ashtekar and M. Bojowald, Class. Quant. Grav. **22**, 3349 (2005), gr-qc/0504029; Class. Quant. Grav. **23**, 391 (2006), gr-qc/0509075; S. D. H. Hsu, Phys. Lett. B **644**, 67 (2007) [hep-th/0608175].
[5] S. D. Mathur, Class. Quant. Grav. **26**, 224001 (2009) [arXiv:0909.1038 [hep-th]].
[6] S. R. Das and A. Jevicki, Mod. Phys. Lett. A **5**, 1639 (1990); D. J. Gross, In *Jerusalem 1990/1991, Proccedings, Two dimensional quantum gravity and random surfaces* 143-173.
[7] C. G. Callan, Jr., S. B. Giddings, J. A. Harvey and A. Strominger, Phys. Rev. D **45**, 1005 (1992) [hep-th/9111056].
[8] M. Natsuume and J. Polchinski, Nucl. Phys. B **424**, 137 (1994) [arXiv:hep-th/9402156]; J. Polchinski, arXiv:hep-th/9411028; J. L. Karczmarek, J. M. Maldacena and A. Strominger, JHEP **0601**, 039 (2006) [arXiv:hep-th/0411174]. S. R. Das and S. D. Mathur, Phys. Lett. B **365**, 79 (1996) [arXiv:hep-th/9507141].
[9] S. D. Mathur, Nucl. Phys. B **529**, 295 (1998) [arXiv:hep-th/9706151].
[10] L. Susskind, arXiv:hep-th/9309145; J. G. Russo and L. Susskind, Nucl. Phys. B **437**, 611 (1995) [arXiv:hep-th/9405117]. A. Sen, Nucl. Phys. B **440**, 421 (1995) [arXiv:hep-th/9411187]; A. Sen, Mod. Phys. Lett. A **10**, 2081 (1995) [arXiv:hep-th/9504147]; C. Vafa, Nucl. Phys. B **463**, 435 (1996) [hep-th/9512078]; G. T. Horowitz and J. Polchinski, Phys. Rev. D **55**, 6189 (1997) [arXiv:hep-th/9612146].
[11] A. Dabholkar, arXiv:hep-th/0409148;
[12] O. Lunin and S. D. Mathur, Nucl. Phys. B **623**, 342 (2002) [arXiv:hep-th/0109154]; O. Lunin, J. M. Maldacena and L. Maoz, hep-th/0212210; I. Kanitscheider, K. Skenderis and M. Taylor, arXiv:0704.0690 [hep-th].
[13] V. S. Rychkov, JHEP **0601**, 063 (2006) [arXiv:hep-th/0512053].
[14] O. Lunin and S. D. Mathur, Phys. Rev. Lett. **88**, 211303 (2002) [arXiv:hep-th/0202072];
[15] A. Strominger and C. Vafa, Phys. Lett. B **379**, 99 (1996) [hep-th/9601029].
[16] S. D. Mathur, A. Saxena and Y. K. Srivastava, Nucl. Phys. B **680**, 415 (2004) [arXiv:hep-th/0311092]; O. Lunin, JHEP **0404**, 054 (2004) [arXiv:hep-th/0404006]; S. Giusto, S. D. Mathur and A. Saxena, Nucl. Phys. B **701**, 357 (2004) [arXiv:hep-th/0405017]; S. Giusto, S. D. Mathur and A. Saxena, Nucl.

Phys. B **710**, 425 (2005) [arXiv:hep-th/0406103]; I. Bena and N. P. Warner, Adv. Theor. Math. Phys. **9**, 667 (2005) [arXiv:hep-th/0408106]; I. Bena and N. P. Warner, Phys. Rev. D **74**, 066001 (2006) [arXiv:hep-th/0505166]; P. Berglund, E. G. Gimon and T. S. Levi, JHEP **0606**, 007 (2006) [arXiv:hep-th/0505167]; A. Saxena, G. Potvin, S. Giusto and A. W. Peet, JHEP **0604**, 010 (2006) [arXiv:hep-th/0509214]; V. Balasubramanian, E. G. Gimon and T. S. Levi, JHEP **0801**, 056 (2008) [arXiv:hep-th/0606118]; J. Ford, S. Giusto and A. Saxena, Nucl. Phys. B **790**, 258 (2008) [arXiv:hep-th/0612227]; I. Bena and N. P. Warner, Lect. Notes Phys. **755**, 1 (2008) [arXiv:hep-th/0701216]; I. Bena, N. Bobev and N. P. Warner, JHEP **0708**, 004 (2007) [arXiv:0705.3641 [hep-th]]; E. G. Gimon and T. S. Levi, JHEP **0804**, 098 (2008) [arXiv:0706.3394 [hep-th]]; I. Bena, C. W. Wang and N. P. Warner, JHEP **0807**, 019 (2008) [arXiv:0706.3786 [hep-th]]; S. Giusto, S. F. Ross and A. Saxena, JHEP **0712**, 065 (2007) [arXiv:0708.3845 [hep-th]]; I. Bena, J. de Boer, M. Shigemori and N. P. Warner, arXiv:1107.2650 [hep-th]; J. de Boer, S. El-Showk, I. Messamah and D. Van den Bleeken, JHEP **0905**, 002 (2009) [arXiv:0807.4556 [hep-th]]; S. Giusto, R. Russo and D. Turton, JHEP **1111**, 062 (2011) [arXiv:1108.6331 [hep-th]]; S. D. Mathur and D. Turton, arXiv:1112.6413 [hep-th]; S. D. Mathur and D. Turton, arXiv:1202.6421 [hep-th].

[17] V. Jejjala, O. Madden, S. F. Ross and G. Titchener, Phys. Rev. D **71**, 124030 (2005) [arXiv:hep-th/0504181].

[18] V. Cardoso, O. J. C. Dias, J. L. Hovdebo and R. C. Myers, Phys. Rev. D **73**, 064031 (2006) [arXiv:hep-th/0512277].

[19] B. D. Chowdhury and S. D. Mathur, Class. Quant. Grav. **25**, 135005 (2008) [arXiv:0711.4817 [hep-th]]; B. D. Chowdhury and S. D. Mathur, Class. Quant. Grav. **25**, 225021 (2008) [arXiv:0806.2309 [hep-th]]; B. D. Chowdhury and S. D. Mathur, Class. Quant. Grav. **26**, 035006 (2009) [arXiv:0810.2951 [hep-th]].

[20] S. D. Mathur, Gen. Rel. Grav. **42**, 113 (2010) [arXiv:0805.3716 [hep-th]].

[21] S. D. Mathur, Int. J. Mod. Phys. D **18**, 2215 (2009) [arXiv:0905.4483 [hep-th]].

[22] D. Marolf, Phys. Rev. D **79**, 044010 (2009) [arXiv:0808.2842 [gr-qc]]; D. Marolf, Phys. Rev. D **79**, 024029 (2009) [arXiv:0808.2845 [gr-qc]].

[23] I. Heemskerk, D. Marolf and J. Polchinski, arXiv:1201.3664 [hep-th].

[24] G. 't Hooft, hep-th/0003004; L. Susskind, J. Math. Phys. **36**, 6377 (1995) [arXiv:hep-th/9409089]; L. Susskind, L. Thorlacius, J. Uglum, Phys. Rev. **D48**, 3743-3761 (1993). [hep-th/9306069]; L. Susskind, Phys. Rev. Lett. **71**, 2367-2368 (1993). [hep-th/9307168]; D. A. Lowe, J. Polchinski, L. Susskind *et al.*, Phys. Rev. **D52**, 6997-7010 (1995). [hep-th/9506138].

[25] S. D. Mathur and C. J. Plumberg, JHEP **1109**, 093 (2011) [arXiv:1101.4899 [hep-th]].

[26] W. Israel, Phys. Lett. A **57**, 107 (1976).

[27] J. M. Maldacena, JHEP **0304**, 021 (2003) [arXiv:hep-th/0106112].

[28] M. Van Raamsdonk, arXiv:0907.2939 [hep-th]; M. Van Raamsdonk, Gen. Rel. Grav. **42**, 2323 (2010) [arXiv:1005.3035 [hep-th]].

Finite 3D de-Sitter quantum Gravity

Rudranil Basu[1], **Samir Ka Paul**[2]

[1,2] S.N. Bose National Centre for Basic Sciences, Sector-III, Block - JD, Salt Lake, Kolkata - 700 098

E-mail: [1] rudranil@bose.res.in, [2] smr@bose.res.in

Abstract. Non-perturbative quantization of 3d gravity is studied in the first order formulation with positive cosmological constant on the background topology of lens space, which is a three sphere modulo a discrete group. We concentrate on the Chern Simons formulation of 3d gravity for the purpose of quantization. However, instead of taking the standard theory we alter it by adding another topological term to the action, which keeps the equations of motion same. This additional term makes the quantization consistent. More importantly the introduction of an new parameter renders the theory finite.

In this short report, which is based on our paper [1], we present how a finite and consistent theory of pure 3D quantum gravity can be developed for the case when the cosmological constant is taken to be positive. Although most of the non-trivial results in 3D gravity are in the context of negative cosmological constant, our work can best be put in the recent upsurge in this arena of de-Sitter [2, 3]. In [2] one-loop partition function of the theory was computed in the functional integral approach. After summing contributions from all possible topologies, the result found was divergent and hence physically meaningless. It was also verified using the Chern Simons formulation too [4]. But in [3], through the introduction of a massive mode in pure gravity through topological massive gravity (TMG) theory, the divergence was shown to be regulated, which comes a bit contrary to one's intuition.

On the other hand people have been looking into the problem of an alternative theory of 3D gravity [6] and its viability for quite some time [7, 8, 10]. Standard theory of Euclidean 3D gravity ($\Lambda > 0$) can be described as a difference between two $SU(2)$ Chern Simons actions (in the units $16\pi G = 1 = c$):

$$\begin{aligned} S &= l\left(I[A^{(+)}] - I[A^{(-)}]\right) \\ \text{where } I[A] &= \int\left(A^I \wedge dA_I + \frac{1}{3}\epsilon_{IJK}A^I \wedge A^J \wedge A^K\right), \qquad A^{(\pm)} = \omega \pm e/l. \end{aligned} \tag{1}$$

Here $l = \dfrac{1}{\sqrt{\Lambda}}$ and e, ω are the triad and connection variables of gravity dynamics. The present author himself and others have worked many features of the alternative theory:

$$\begin{aligned} \tilde{S}[A^{(+)}, A^{(-)}] &= \frac{k_{(+)}}{2\pi}I[A^{(+)}] + \frac{k_{(-)}}{2\pi}I[A^{(-)}] \\ \text{where} \quad k_{(\pm)} &= \frac{l\,(1/\gamma \pm 1)}{8G}, \quad \text{reinstating } G \end{aligned} \tag{2}$$

Hereby we have introduced the dimensionless parameter γ. It is a feasible theory because equations of motion coming from (2) are same as those coming from (1). The only difference is that the canonical structure of this augmented theory is different from that of the standard theory. In absence of boundaries of the base manifold, the pre-symplectic structure looks as:

$$\Omega\left(\delta_1, \delta_2\right) = \frac{k_{(+)}}{\pi}\int_\Sigma \delta_1 A^{(+)} \wedge \delta_2 A^{(+)} + \frac{k_{(-)}}{\pi}\int_\Sigma \delta_1 A^{(-)} \wedge \delta_2 A^{(-)}. \tag{3}$$

In order to proceed for quantization of the theory, we note that, topology of the background manifold is the sole deciding factor about the quantum theory, since this is a topological field theory. Along with it, we remind ourselves that the vacuum solution of $\Lambda > 0$ gravity is de-Sitter space. When Euclideanized, this is dS $\rightarrow S^3$ and all solutions are locally isomorphic to S^3, geometrically. We take for viable topology S^3/Γ, where Γ is a discrete group. For example, if we choose $\Gamma \sim \mathbb{Z}_p$ and specify its action on S^3 in a particular way (involving another natural number $q \leq p$ and $(q,p) = 1$), we get Lens spaces, L(p,q). There are other choices for Γ, whose contributions are trivial.

One may now be optimist enough to look forward to the possibility of quantizing the theory directly (geometric quantization [5]). But this is the point where the background come in the way. Gauge invariant physical phase space of Chern Simons theory is given by : $\hom(\pi_1(M) \rightarrow G)/\sim$. Here M is the base space and G is the gauge group. $\sim$ here denotes gauge equivalence. In this case we can easily see that the gage invariant phase space consists of a total of p^2 points, hence cannot qualify for symplectic manifold. Obviously then we take a less direct route and follow [4].

In this later approach one contemplates $L(p,q)$ to be constructed by gluing together 2 solid tori at their surfaces through an element of the mapping class group $PSL(2,\mathbb{Z}$ of the toric surface. In the defining representation, an element U of the group, which generates $L(p,q)$ is

$$U = \begin{pmatrix} q & b \\ p & d \end{pmatrix} \in SL(2,\mathbb{Z}). \tag{4}$$

This U can be generated by the well-known S and T modular-transformation operators as

$$U = S \prod_{s=1}^{t-1} (T^{m_s} S), \qquad m_s \in \mathbb{Z}^+.$$

The following restrictions of m_s are to be followed:

$$p/q = -m_{t-1} + \cfrac{1}{m_{t-2} - \cfrac{1}{\dots - \cfrac{1}{m_1}}}. \tag{5}$$

Meanwhile, one has to keep the quantization ready on the toric surfaces ready. This is easy and can be carried out in the framework of geometric quantization. Without going into the detail, we state that after performing the quantization we end up with a Hilbert space associated with each toric surface for each Chern Simons theory with label k. The dimensionality of the Hilbert space over the field $\mathbb{C}$ is $k+1$. A convenient basis, for which the inner product structure is easily spelled out is given in terms of the $SU(2)$ theta functions:

$$\psi_{j,k}(z,\tau) = \frac{\vartheta_{j+1,k+1}(z,\tau) - \vartheta_{-j-1,k+2}(z,\tau)}{\vartheta_{1,2}(z,\tau) - \vartheta_{-1,2}(z,\tau)} \qquad j = 0, \dots k \tag{6}$$

Here $z = \frac{1}{\pi}(\theta + \tau\phi)$ coordinatizes the phase space with $\theta, \phi \in [0, 2\pi)$, τ being the modular parameter on the phase space, which is a Kähaler manifold and also a Jacobian variety. This Hilbert space on the other hand makes sense if and only if:

$$a := \frac{l}{8l_p} = s/2 \quad s \in \mathbb{N} \quad \text{and} \quad \gamma = \frac{a}{(a-1)+t} \quad t \in \mathbb{N}. \tag{7}$$

$l_p = G\hbar/c^3$ is the Planck length. The axiomatics of topological field theory [9] now dictates that the partition function of the glued space (for each Chern Simons theory), ie the Lens space $L(p,q)$ must be given by

$$Z(r)_{L(p,q)} = \langle \psi_{0,k} | U | \psi_{0,k} \rangle.$$

In order to evaluate this, one keeps in mind the actions of the S and T transformations on the space of the theta functions. A non-trivial calculation in this direction gives us [4]:

$$Z(r)_{L(p,q)} = -\frac{i}{\sqrt{2rp}} \exp\left(6\pi i s(q,p)/r\right) \sum_{\pm} \sum_{n=1}^{p} \exp\left(\frac{2\pi i q r n^2}{p} + \frac{2\pi i n(q \pm 1)}{p} \pm \frac{\pi i}{rp}\right), \tag{8}$$

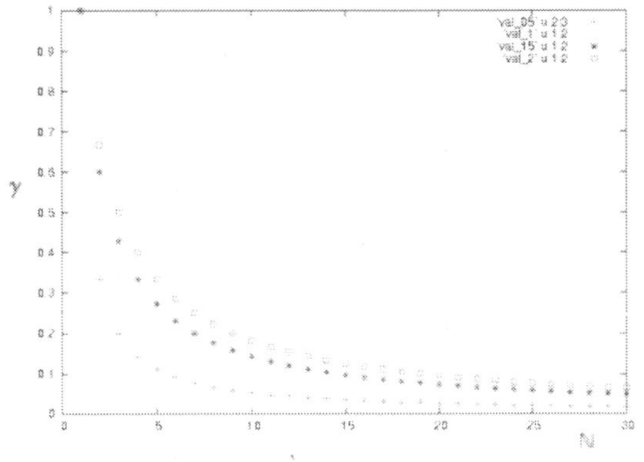

Figure 1. Some of the first allowed values of γ for $l/l_p = 4, 8, 12, 16$

denoting $r = k + 2$. Here $s(q,p) = \sum_{l=1}^{p-1} \frac{l}{p}\left(\frac{lq}{p} - \left[\frac{lq}{p}\right] - \frac{1}{2}\right)$ is the Dedekind sum. [] is the standard floor function.

Returning to our original problem of gravity, we have the total partition function, summing over all topologies:

$$Z^{\text{tot}} = \sum_{p=1}^{\infty} \sum_{\substack{q(\text{mod}p)\\(q,p)=1}} = Z(r_{(+)})_{L(p,q)} Z(r_{(+)})_{L(p,q)} \tag{9}$$

For the purpose of illuminating the convergence of our theory we choose a pair of particular saddles in the sum (8), $n = \frac{q \pm 1}{2}$ for the '+' and the '-' type Chern Simons theory respectively. Moreover we relax the above restrictions (7) a little bit by demanding $a/\gamma \in 2\mathbb{N}$. Provided these conditions are obeyed, we re-write (9):

$$\begin{aligned} Z^{\text{tot}} &= -\frac{1}{2\sqrt{r_{(+)}r_{(-)}}} \sum_{p=1}^{\infty} \frac{1}{p} \sum_{\substack{q(\text{mod}p)\\(q,p)=1}} \exp\left(6\pi i s(q,p)/R_+\right) \exp\left(\frac{\pi i}{p}\left(2a + (q+q^*)\left(a/\gamma + 2\right)\right)\right) \times \\ & \left[e^{\frac{\pi i}{pR_+} + \frac{2\pi i}{p}(q+1)} + e^{-\frac{\pi i}{pR_+} + \frac{2\pi i}{p}(q-1)} - e^{\frac{\pi i}{pR_-} + \frac{4\pi i}{p}} - e^{\frac{-\pi i}{pR_-}} \right] \end{aligned} \tag{10}$$

where $\frac{1}{R_\pm} = \frac{1}{r_{(+)}} \pm \frac{1}{r_{(-)}}$

This is the point where one encounters the divergence found in [2] through $\sum_{\substack{q(\text{mod}p)\\(q,p)=1}} 1 = \phi(p)$, the Euler totient function. But in our case, in the above expression, there is no q independent term. This is the reason we can avoid the dangerous totient function. For the sake of investigation of convergence we further go to the limit: $0 < \gamma \ll 1$. This limit is consistent with the restrictions (7). In this limit,

$$Z^{\text{tot}} = -\frac{\gamma}{a}\left(1 - \frac{2\gamma}{a}\right) \sum_{p=1}^{\infty} \frac{1}{p} e^{\frac{2\pi i a}{p}} \cos(2\pi/p) \left[S\left(\frac{a}{2\gamma} + 2, \frac{a}{2\gamma} + 1; p\right) - e^{\frac{2\pi i}{p}} S\left(\frac{a}{2\gamma} + 1, \frac{a}{2\gamma} + 1; p\right) \right] \tag{11}$$

$$S(\alpha, \beta; p) = \sum_{\substack{q(\text{mod}p)\\(q,p)=1}} \exp\left(2\pi i(\alpha q + \beta q^*)/p\right).$$

When one expands the cosine and the exponential above in powers of π/p, Kloostermann zeta functions:

$$L(m, n; s) = \sum_{p=1}^{\infty} p^{-2s} S(m, n; p)$$

are evident. These functions are analytic in $\Re s > 1/2$. In the small γ region, the above summand of (11) becomes:

$$\sum_{m,n,r=0}^{\infty} \frac{(2\pi i)^{r+n+2m+1}}{r+1} \frac{a^n}{(2m)!n!r!} L(\frac{a}{2\gamma}, \frac{a}{2\gamma}; \frac{r+n+2m}{2}+1) \quad (12)$$

As we learnt above, all the above terms are analytic. Hence the partition function is free from any divergence. This is our main result.

Now we go on to present a brief comparative study between our theory and TMG. Denoting by E, the contributions coming from pure Einstein Hilbert theory with cosmological constant and by MG, the ones coming from massive graviton modes, we have the one-loop(in contrast to our non-perturbative result) result [3]:

$$\sum_{p=1}^{\infty} \sum_{\substack{q(\text{mod}p) \\ (q,p)=1}} Z_{\text{E}}^{(0)} Z_{\text{MG}}^{(0)} Z_{\text{E}}^{(1)} \sim \sum_{r=0}^{\infty} \frac{(2\pi a)^r}{r!} L(\frac{a}{2\gamma}, \frac{a}{2\gamma}; \frac{r}{2}+\frac{1}{2}) + \text{trivially analytic terms.} \quad (13)$$

Comparing this with (12), the interesting fact tht crops up is that the term corresponding to $r = 0$ in the sum of the RHS is the source of divergence since it corresponds to Kloosterman zeta function with $s = 1/2$. But in [3] it is showed that when one includes $Z_{MG}^{(1)}$ as the product and then performs the sum over p, the divergence goes away. Up to one loop calculation they have

$$Z = \sum_{p=1}^{\infty} \sum_{\substack{q(\text{mod}p) \\ (q,p)=1}} Z_{\text{E}}^{(0)} Z_{\text{MG}}^{(0)} Z_{\text{E}}^{(1)} Z_{\text{MG}}^{(1)}.$$

The expression of $Z_{\text{MG}}^{(1)}$ as given in [3] is far too complicated for the above expression to be analytically simplified and compared with (12). The way the divergence in the above expression is controlled by $Z_{\text{MG}}^{(1)}$ is very like the way we showed (12) to be finite. We conclude that although the finiteness of TMG could be ascribed to its propagating graviton modes , our theory (2), being devoid of massive gravitons still yield a reasonably similar convergent partition function.

Acknowledgments
RB thanks Council for Scientific and Industrial Research (CSIR), India, for support through the SPM Fellowship SPM-07/575(0061)/2009-EMR-I. The authors also thank the anonymous referee for comments which helped improving the manuscript considerably.

[1] R. Basu and S. KPaul, Phys. Rev. D **85**, 023520 (2012) [arXiv:1109.0793 [hep-th]].
[2] A. Castro, N. Lashkari, A. Maloney, "A de Sitter Farey Tail," [arXiv:1103.4620 [hep-th]].
[3] A. Castro, N. Lashkari, A. Maloney, "Quantum Topologically Massive Gravity in de Sitter Space," [arXiv:1105.4733 [hep-th]].
[4] L. C. Jeffrey, Commun. Math. Phys. **147**, 563-604 (1992).
[5] S. Axelrod, S. Della Pietra, E. Witten, J. Diff. Geom. **33**, 787-902 (1991).
[6] E. Witten, Nucl. Phys. **B311**, 46 (1988).
[7] R. Basu, S. KPaul, Class. Quant. Grav. **27**, 125003 (2010). [arXiv:0909.4238 [gr-qc]].
[8] V. Bonzom, E. R. Livine, Class. Quant. Grav. **25**, 195024 (2008). [arXiv:0801.4241 [gr-qc]].
[9] E. Witten, Commun. Math. Phys. **121**, 351 (1989).
[10] R. Basu, A. Chatterjee, Class. Quant. Grav. **28**, 225013 (2011). [arXiv:1101.2724 [gr-qc]].

Study of Thermodynamic Quantities in Horava-Lifshitz and $f(R)$ gravity Theories

Samarpita Bhattacharya[1], Ujjal Debnath[2], Surajit Chattopadhyay[3]

[1] Research Scholar, Department of Mathematics, Bengal Engineering and Science University, Shibpur, Howrah-711 103, India

E-mail: samarpita_sarbajna@yahoo.co.in

[2] Assistant Professor, Department of Mathematics, Bengal Engineering and Science University, Shibpur, Howrah-711 103, India

E-mail: ujjaldebnath@yahoo.com , ujjal@iucaa.ernet.in

[3] Assistant Professor, Department of Computer Application, Pailan College of Management and Technology, Kolkata 700 104, India

E-mail: surajcha@iucaa.ernet.in , surajit_2008@yahoo.co.in

Abstract. The present paper reports studies on thermodynamic quantities like temperature of the universe, heat capacity and squared speed of sound in $f(R)$ and Horava-Lifshitz gravity theories. Considering the universe to be filled with dark matter and dark energy we have shown that in all cases the equation of state behaves like quintessence. The thermodynamic quantities have been studied graphically by plotting them against redshift z.

1. Introduction

Accelerated expansion of the universe is well documented in literature [1]. There are two classes of models to explain this accelerated expansion. The classes are: (i) Dark energy and (ii)Modified gravity. In the first class, a new matter component characterized by negative pressure is invoked and is dubbed as "dark energy". This is described by an equation of state $w = p/\rho$, namely the ratio of the homogeneous dark energy pressure over the energy density. For a review of dark energy see [1]. In the second case, one modifies the laws of gravity whereby a late-time acceleration is produced [2]. Different models of modified gravity have been suggested till date. A review on modified gravity theory is available in [3]. In the present work, two types of modified gravity are considered. They are $f(R)$ gravity [4, 5] and Horava-Lifshitz gravity [6, 7]. The $f(R)$ gravity has the interesting feature that choosing an observationally reliable $f(R)$, it is possible to describe the early inflation as well as the late time acceleration of the universe in a fascinating manner [8]. The basic idea of Horava-Lifshitz (HL) gravity is to modify the UV behavior of the general theory so that the theory is perturbatively renormalizable [9]. The action of the HL gravity is given by [10]

$$\begin{array}{c} I = dt\int dtd^3x(\mathcal{L}_0 + \mathcal{L}_1 + \mathcal{L}_m); \;\; \mathcal{L}_0 = \sqrt{g}N\left[\frac{2}{\kappa^2}(K_{ij}K^{ij} - \lambda K^2) + \frac{\kappa^2\mu^2(\Lambda R - 3\Lambda^2)}{8(1-3\lambda)}\right] \\ \mathcal{L}_1 = \sqrt{g}N\left[\frac{\kappa^2\mu^2(1-4\lambda)}{32(1-3\lambda)}R^2 - \frac{\kappa^2}{2\omega^4}(C_{ij} - \frac{\mu\omega^2}{2}R_{ij})(C^{ij} - \frac{\mu\omega^2}{2}R^{ij})\right] \end{array} \quad (1)$$

Journal of Physics: Conference Series **405** (2012) 012007 doi:10.1088/1742-6596/405/1/012007

where, κ^2, λ, μ, ω and Λ are constant parameters, and C_{ij} is Cotton tensor (conserved and traceless, vanishing for conformally flat metrics). $\mathcal{L}_m$ stands for the Lagrangian of other matter field. Modified field equations for this case are thoroughly explained in the reference [10].

The action of $f(R)$ gravity is given by [11]

$$S = \int d^4x\sqrt{-g}\left[\frac{f(R)}{2\kappa^2} + \mathcal{L}_{matter}\right] \tag{2}$$

where g is the determinant of the metric tensor $g_{\mu\nu}$, $\mathcal{L}_{matter}$ is the matter Lagrangian and $\kappa^2 = 8\pi G$. The $f(R)$ is a non-linear function of the Ricci curvature R that incorporates corrections to the Einstein-Hilbert action which is instead described by a linear function $f(R)$. The gravitational field equations are elaborated in the reference [12].

2. Thermodynamic quantities

We consider the FRW universe treated as a thermodynamical system. Then from Gibb's equation of thermodynamics, we have [13] $TdS = d(\rho V) + pdV = d((\rho + p)V) - Vdp$, where S is the entropy, T is the temperature and V is the volume of the universe. The integrability condition of thermodynamic system is given by [13] $\frac{\partial^2 S}{\partial T \partial V} = \frac{\partial^2 S}{\partial V \partial T}$. Subsequently, we get $S = \frac{(\rho+p)V}{T}$. However, for adiabatic process entropy is constant and consequently $d[(\rho + p)] = Vdp$. The expressions for temperature T, squared speed of sound v_s^2 and heat capacity C_V can be obtained as $T = \frac{(\rho+p)V}{S}$; $v_s^2 = \frac{\partial p}{\partial \rho}$; $C_V = V\frac{\partial \rho}{\partial T}$ (see ref. [14] for details).

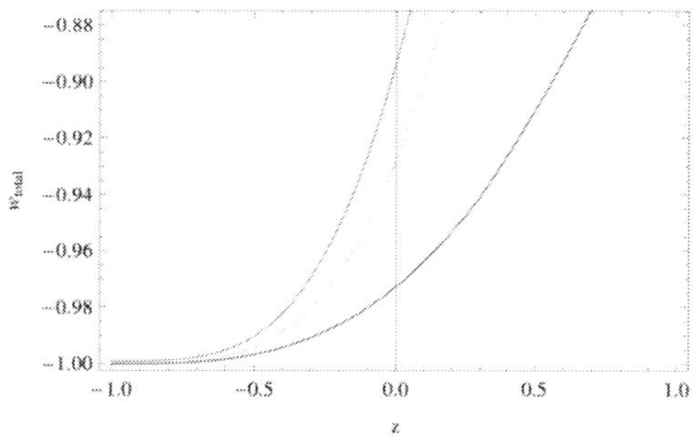

Figure 1. This figure shows the evolution of w_{total} with evolution of the universe for HL gravity.

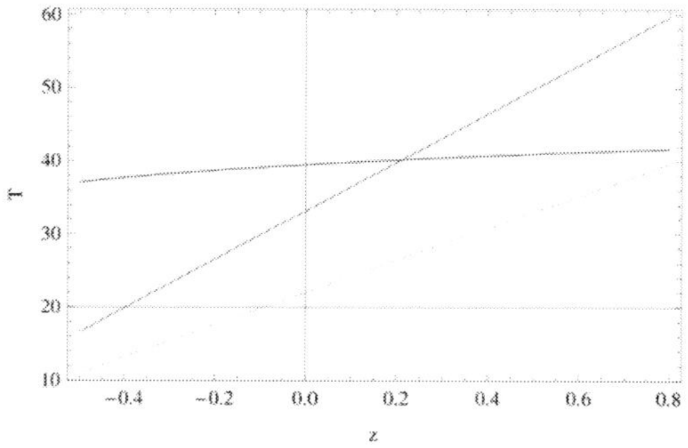

Figure 2. This figure shows the behavour of T with evolution of the universe for HL gravity.

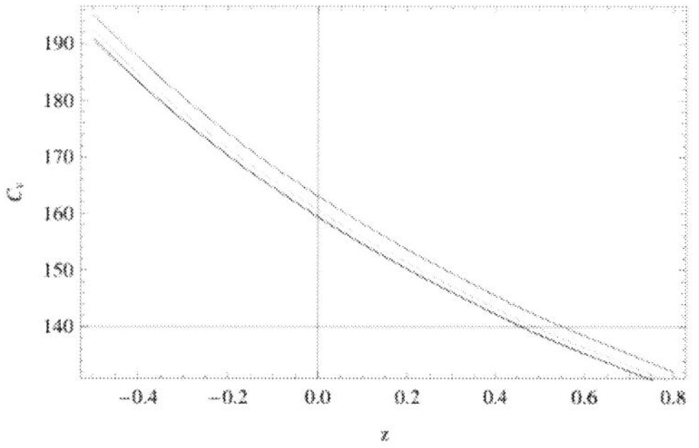

Figure 3. This figure shows the behavior of C_v with evolution of the universe for HL gravity.

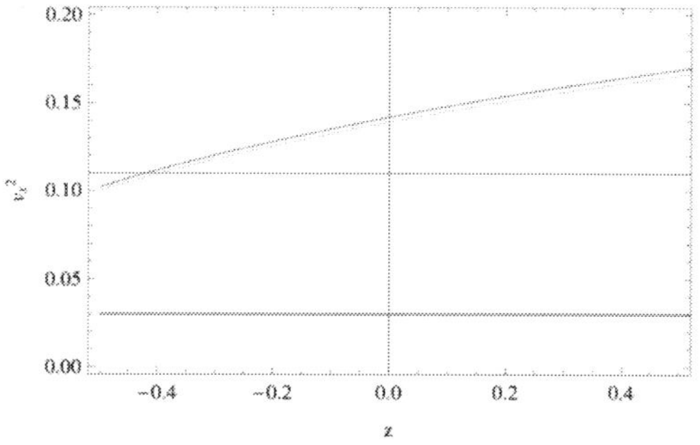

Figure 4. This figure shows v_s^2 with evolution of the universe for HL gravity.

Journal of Physics: Conference Series **405** (2012) 012007 doi:10.1088/1742-6596/405/1/012007

2.1. Thermodynamic quantities under Horava-Lifshitz gravity

Under HL gravity, the total energy density and total pressure are expressed in terms of redshift ($z = \frac{1}{a} - 1$) as

$$\rho(z) = \rho_m + \rho_D = \rho_{m_0}(1+z)^{3(1+w_m)} + \frac{1}{16\pi G_c}\left(\frac{3k^2(1+z)^4}{\Lambda} + 3\Lambda\right) \tag{3}$$

$$p(z) = p_m + p_D = \rho_{m_0} w_m (1+z)^{3(1+w_m)} + \frac{1}{16\pi G_c}\left(\frac{k^2(1+z)^4}{\Lambda} - 3\Lambda\right) \tag{4}$$

Where, $G_c = \frac{\kappa^2 c}{16\pi(3\lambda-1)}$ and $c = \frac{\kappa^2\mu}{4}\sqrt{\frac{\Lambda}{1-3\lambda}}$. Subsequently, the thermodynamic quantities are expressed in terms of redshift z as

$$v_s^2(z) = \frac{k^2(1+z)\kappa^2\mu^2 + 6(1+z)^{3w_m}(-1+3\lambda)w_m\rho_{m0}(1+w_m)}{3\{k^2(1+z)\kappa^2\mu^2 + 2(1+z)^{3w_m}(-1+3\lambda)(1+w_m)\rho_{m0}\}} \tag{5}$$

and

$$C_v(z) = \frac{3S_0\{k^2(1+z)\kappa^2\mu^2 + 2(1+z)^{3w_m}(-1+3\lambda)(1+w_m)\rho_{m0}\}}{k^2(1+z)\kappa^2\mu^2 + 6(1+z)^{3w_m}(-1+3\lambda)w_m\rho_{m0}(1+w_m)} \tag{6}$$

Using the expressions of thermodynamic quantities expressed in the earlier section we can get T and w_{total} in terms of z. The thermodynamic quantities obtained above are plotted against z in figures 1 to 4. In the figures, we make the plots for $k = -1$ (red), 1 (green) as well as 0 (blue).

2.2. Thermodynamic quantities under $f(R)$ gravity

In the present section, while considering the $f(R)$ gravity, we have illustrated with a solution $f(R) = \beta R + \alpha R^m$, $R = \frac{A}{a^n}$, $m > 1$, $\alpha > 0$, $\beta > 0$. Using the above form of $f(R)$ in we have computed ρ_1, p_1, ρ_c and p_c in terms of z as

$$\rho_1 = \frac{(1+z)^{3(1+w_m)}\rho_{m0}}{m(A(1+z)^n)^{-1+m}\alpha + \beta};\quad p_1 = \frac{w_m(1+z)^{3(1+w_m)}\rho_{m0}}{m(A(1+z)^n)^{-1+m}\alpha + \beta} \tag{7}$$

$$\rho_c = \frac{\alpha(-1+m)A^m(1+z)^{mn}}{16\pi(mA^{-1+m}(1+z)^{n(-1+m)}\alpha + \beta)}\left\{1 + \frac{mn(2m-4)(1+z)^{4-2n}\left(3C_1 - \frac{3k}{(1+z)^2} + \frac{A(1+z)^{-4+n}}{n-4}\right)}{A^2}\right\} \tag{8}$$

$$p_c = \frac{A(1+z)^{2n}}{48m\pi}\{-3A + 6(-2+m)mn^2(1+n)^2(1+z)^{4-n} + +6mn(1+z)^{2-n}(-5+n+z+nz)\sqrt{C_1 - \frac{k}{(1+z)^2} + \frac{A(1+z)^{-4+n}}{3(-4+n)}}\} \tag{9}$$

Using ρ_1, p_1, ρ_c and p_c as above we have derived the equation of state parameter and the thermodynamic quantities and plotted them in figures 5 to 8.

3. Concluding remarks

In this study, we have considered Horava-Lifshitz and $f(R)$ gravity theories. In each case we have considered that the universe is filled with dark matter and dark energy which are not interacting. Prior to discussing the thermodynamic quantities we have studied the behaviors of the equation of state parameters. In both the cases the equation of state parameter is found to stay above -1 that indicates quintessence like behavior. In both of the cases, the C_v remains at the positive level throughout the evolution of the universe. Temperature T is always found to fall with evolution of the universe. It is known [15] that deviation from $v_s^2 = 1$ is a direct sign that the dark energy is a complex, dynamical fluid rather than an inert cosmological constant. In the case of $f(R)$, it stays near 1 in the early stage of universe. However, at late stages, it stays much above 1. A negative heat capacity is impossible for any system in stable equilibrium [16]. The positive heat capacity in all of the gravities indicate equilibrium.

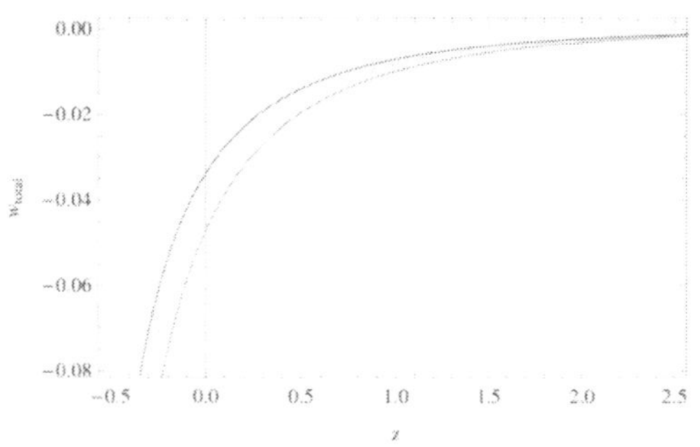

Figure 5. This figure shows the evolution of w_{total} with evolution of the universe for $f(R)$ gravity.

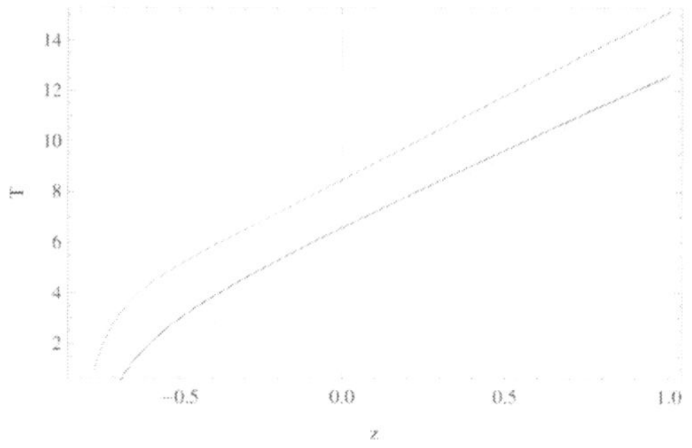

Figure 6. This figure shows the behaviour of T with evolution of the universe for $f(R)$ gravity.

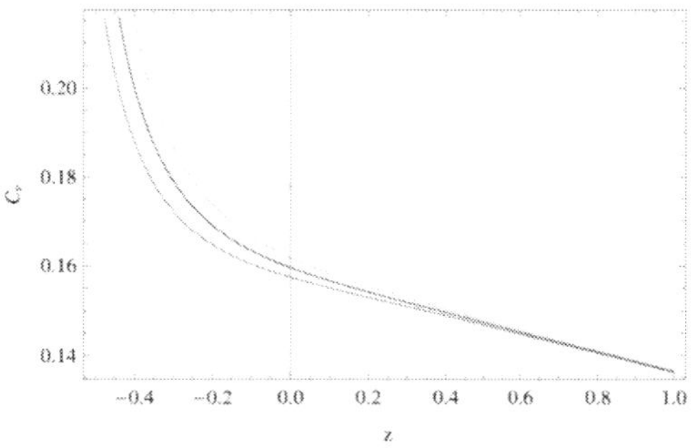

Figure 7. This figure shows the behavior of C_v with evolution of the universe for $f(R)$ gravity.

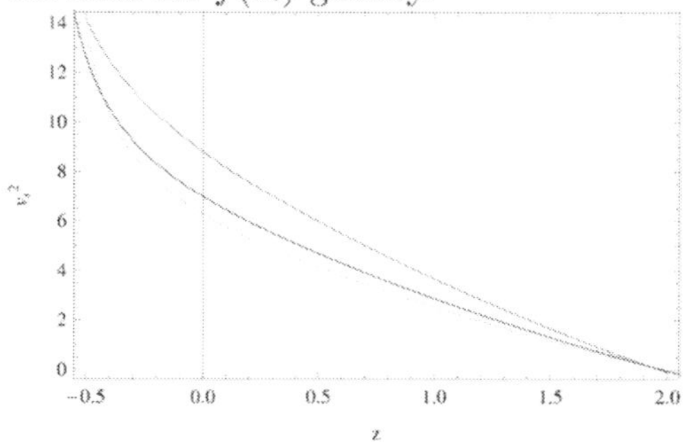

Figure 8. This figure shows v_s^2 with evolution of the universe for $f(R)$ gravity.

Acknowledgement
Sincere thanks are due to the organizers of COSGRAV 2012 for kindly giving us the opportunity to present this work.

References
[1] Copeland EJ, Sami M and Tsujikawa S 2006 *Int. J. Mod. Phys. D* **15** 1753
[2] Carloni S, Chaichian M, Nojiri S, Odintsov S D, Oksanen M and Tureanu A 2010 *Phys. Rev. D* **82** 065020
[3] Nojiri S and Odintsov SD 2007 *Int. J. Geom. Meth. Mod. Phys.* **4** 115
[4] Carroll S M, Duvvuri V, Trodden M and Turner M S 2004 *Phys. Rev. D* **70** 043528
[5] Sotiriou TP and Faraoni V 2010 *Rev. Mod. Phys.* **82** 451
[6] Vernieri D and Sotiriou TP 2012 *Phys. Rev. D* **85** 064003
[7] Sotiriou TP 2011 *J. Phys. Conf. Ser.* **283** 012034
[8] Faraoni V 2008 *Phy. Lett. B* **665** 135
[9] Kluson J 2010 *JHEP* **07** 038
[10] Jamil M, Saridakis EN and Setare MR 2010 *JCAP* **11** 032
[11] Bamba K and Geng CQ 2009 *Phys. Lett. B* **679** 282
[12] Kiritsis E and Kofinas G 2009 *Nucl. Phys. B* **821** 467
[13] Izquierdo G and Pavon D 2006 *Phys. Lett. B* **639** 1
[14] Chattopadhyay S, Debnath U and Bhattacharya S 2012 *Int. J. Theor. Phys.* DOI 10.1007/s10773-012-1198-3
[15] Steinhardt PJ 2003 *Phil. Trans. R. Soc. Lond. A* **361** 2497.
[16] Mazur PO and Mottola E 2004 *Proceedings of the National Academy of Sciences (PNAS)* **101** 9545.

Increase of entropy for black holes in Einstein-Gauss-Bonnet theory[1]

Ayan Chatterjee[1 2]

[1] The Institute of Mathematical Science, C.I.T. Campus, Taramani, Chennai-600113

E-mail: `ayan@tifr.res.in`

Abstract. We establish the physical process version of first law by studying small perturbations of a stationary black hole with regular bifurcation surface in Einstein-Gauss-Bonnet (EGB) gravity. We shaow that when a stationary black hole is perturbed by matter stress energy tensor and settles down to a new stationary state, the Wald entropy increases if the matter satisfies null energy condition.

The striking similarity of the laws of black hole mechanics with thermodynamics was first established in case of general relativity (GR) [1]. It is interesting to explore whether this analogy is a peculiar property of GR or a robust feature of any generally covariant theory of gravity.
To begin, zero'th law, which ensures the constancy of surface gravity is valid for any stationary Killing horizon with a regular bifurcation surface [2], irrespective of the gravitational dynamics. The equilibrium state version of first law is established by Wald and collaborators [3] for any arbitrary diffeomorphisom invariant theory of gravity. For black holes in GR, the analog of second law of thermodynamics is the "area theorem" which asserts that area of a black hole can not decrease in any classical process [4]. A simple illustration of this result can be obtained by studying the physical process version of the first law which describes the dynamical change of horizon area in response to a flux of matter through the horizon [5]. In case of stationary bifurcate horizons with a horizon generating Killing field ξ^a, the physical process first law is,

$$\Delta A = -\frac{8\pi}{\kappa}\int_{\mathcal{H}} T_{ab}\xi^a dH^b \tag{1}$$

where κ is the surface gravity of the black hole, T_{ab} is the stress energy tensor of the matter and the integration is over the horizon $\mathcal{H}$. The measure is $dH_b = -k_b\, d\lambda\sqrt{\sigma}\, dA$, where λ is an affine parameter along the horizon generators, $k^b = (\partial_\lambda)^b$ is tangent to the generators, and $\sqrt{\sigma}dA$ is the area element of a constant λ horizon slice. Eq.(1) shows that in a dynamical process, the area of a black hole can not decrease as long as the matter satisfies null energy condition (i.e $T_{ab}k^a k^b \geq 0$). Once we identify the Hawking temperature as $T = \kappa/2\pi$, Eq.(1) becomes completely analogous to the Clausius relationship $\Delta S = \Delta Q/T$ as in ordinary thermodynamic systems.
The question of validity of the second law for arbitrary theory of gravity still remains an unresolved issue. We shall look for a physical process version of first law beyond GR to establish

[1] This talk is based on a work done with S. Sarkar

[2] Present address: Tata Institute of Fundamental Research, Homi Bhabha Road, Colaba, Mumbai-400005.

that the entropy can not decrease at least for dynamical process which takes the black hole from a initial stationary state to a final one. If the physical process and the equilibrium state versions of the first law do not match, this will suggest an inconsistency in the assumptions behind the derivation, in particular to the assumptions that the black hole horizon is not destroyed by throwing matter into it and the final state in the late time is again stationary. The fact that in GR, both these versions agree, lends strong support to the idea that gravitational collapse result in a predictable black hole. It is expected that black holes in any reasonable theory of gravity must have such property.

We provide a direct proof of the physical process version of the first law for Einstein-Gauss-Bonnet (EGB) gravity [6]. Our proof involves the direct evaluation of the change of the Wald entropy when a stationary black hole is weakly perturbed and finally settles down to a new stationary state. We show that in such a process, the Wald entropy for EGB theory can not decrease as long as the matter stress energy tensor satisfies null energy condition. This firmly establishes that the Wald entropy for stationary black holes merits the name entropy even for more general gravity theories. Let us begin with properties of a stationary black hole horizon[3]. In a D-dimensional spacetime, the event horizon is a null hyper-surface $\mathcal{H}$ parametrized by an affine parameter λ. The vector field $k^a = (\partial_\lambda)^a$ is tangent to the horizon and obeys geodesic equation. Consider a spatial cross section at $\lambda = 0$ everywhere. All such $\lambda =$ constant slices are space-like and foliate the horizon. Any point p on such slices have coordinates $\{\lambda, x^A\}$ where x^A, $(A = 2, \cdots, D)$ are the coordinates of a point on $\lambda = 0$ slice connected with p by a horizon generator. We can construct a basis with the vector fields, $\{k^a, l^a, e^a_A\}$ where l^a is a second null vector such that $l^a k_a = -1$. The induced metric on any slice is $\gamma_{ab} = g_{ab} + 2k_{(a}l_{b)}$. The change of the induced metric from one slice to another can be obtained from the metric evolution equation [8],

$$\mathcal{L}_k \gamma_{ab} = 2\left(\sigma_{ab} + \frac{\theta}{(D-2)}\gamma_{ab}\right), \tag{2}$$

where σ_{ab} is the shear and θ is the expansion of the horizon. If the event horizon is also a Killing horizon [4], i.e. the horizon generators are the orbits of a Killing field $\xi^a = (\partial/\partial v)^a$, which is null on the horizon, we can define the surface gravity κ by the relationship $\xi^a \nabla_a \xi^b = \kappa\, \xi^b$. For stationary spacetimes with a Killing horizon, both the expansion and shear vanish and using Raychaudhuri equation and the evolution equation for shear, we obtain [8], $R_{ab}k^a k^b = k^a k^c \gamma^b_m \gamma^d_n C_{abcd} = 0$ where C_{abcd} is the Weyl tensor.

The Einstein tensor (and hence the Ricci tensor) takes a highly symmetric, block diagonal form on a stationary Killing horizon with a regular bifurcation surface [9]. To understand this result, let us construct a basis $\{\xi^a, N^a, e^a_A\}$, where N^a is another null vector satisfying $\xi^a N_a = -1$ and express Ricci tensor in this basis. On the stationary Killing horizon, Ricci tensor satisfies the relation $R_{ab}\xi^b \sim \xi_a$ [8]. Using this, it is straightforward to show, that on the bifurcation surface, the Ricci tensor must have a form, $R_{ab} = C_1 g^{\perp}_{ab} + C_2 \gamma_{ab}$, where $g^{\perp}_{ab} = -2k_{(a}l_{b)}$, is the metric of the 2-dimensional plane orthogonal to the $(D-2)$-dimensional bifurcation surface. Since Ricci tensor is a Killing invariant, we can Lie propagate the Ricci tensor onto any other space-like section of the horizon. The coefficients C_1 and C_2 are given by, $C_1 = -R_{ab}k^a l^b$ and $C_2 = R_{ab}\gamma^{ab}/(D-2)$. For stationary Killing horizons, since both the expansion and shear vanish, using the block diagonal form of the Ricci tensor on the horizon, it is also possible to show that [10], $k^a \gamma^b_i \gamma^c_j \gamma^d_k C_{abcd} = 0$. Another relationship which will be useful for our calculation

[3] We adopt the metric signature $(-,+,+,+,...)$ and our sign conventions are same as those of [8].

[4] Here we make an implicit assumption, that the event horizon of a stationary black hole is also a Killing horizon with regular bifurcation surface. Although this is certainly true for GR, we are not aware of any proof for EGB gravity.

is the expression of intrinsic Ricci curvature ${}^{(D-2)}R^{ab}$ of the cross-section of the stationary Killing horizon in terms of full curvatures, given by [11]

$$ {}^{(D-2)}R_{ab} = \gamma_a^m \, \gamma_b^n \, \gamma_s^r \, R^s_{mrn}. \tag{3} $$

Next, we discuss the Gauss-Bonnet (GB) term, $\mathcal{L}_{GB} = R^2 - 4R_{ab}R^{ab} + R_{abcd}R^{abcd}$ in spacetime dimensions $D > 4$. EGB gravity is free from ghosts, leads to a well-defined initial value problem and therefore, is a reasonable candidate for a low energy effective theory of gravity. The action functional is given by, $\mathcal{L} = \frac{1}{16\pi}\int d^Dx\sqrt{-g}\ (R + \alpha\mathcal{L}_{GB})$. The field equation of EGB theory is, $G_{ab} + \alpha H_{ab} = 8\pi T_{ab}$ where, $H_{ab} \equiv 2\Big[RR_{ab} - 2R_{aj}R^j{}_b - 2R^{ij}R_{aibj} + R_a{}^{ijk}R_{bijk}\Big] - \frac{1}{2}g_{ab}\mathcal{L}_{GB}$. For the EGB action, the Wald entropy associated with a stationary Killing horizon is [12] ,

$$ S = \frac{1}{4}\int \rho\, \sqrt{\gamma}\, dA, \tag{4} $$

where the entropy density $\rho = \left(1 + 2\alpha\ {}^{(D-2)}R\right)$ and the integration is over $(D-2)$-dimensional space-like cross-section of the horizon. ${}^{(D-2)}R$ is the intrinsic Ricci scalar of the horizon cross-section. Since the black hole is stationary in the asymptotic future, the vector field ξ^a is an exact Killing vector at late times and all the Lie derivatives of dynamical fields w.r.t ξ^a vanish. The accretion process is assumed to be slow such that all changes of the dynamical fields are first order and we can neglect all viscous effects. More specifically, we assume that: $\theta \sim \sigma_{ab} \sim \mathcal{O}(\epsilon)$. The bifurcation surface at $\lambda = 0$ is taken as the initial cross-section and the final cross-section is in the asymptotic stationary regime. Then the change is entropy is,

$$ \Delta S \;=\; \int_{\mathcal{H}} \lambda\left(2\pi\, T_{ab} - \frac{\alpha}{8\pi}\mathcal{R}_{ab}\right)\, k^a k^b d\lambda\, \sqrt{\gamma}\, dA, \tag{5} $$

In deriving the previous equation, a total derivative is discarded since it vanishes both on the initial $\lambda = 0$ slice and on the final stationary state. Also, we have used the Raychaudhuri equation and neglected all terms except those first order in perturbation. We have also used the field equations and the fact that $\mathcal{R}_{ab} = H_{ab} - 2^{(D-2)}RR_{ab} + 2\nabla_a\nabla_b\, {}^{(D-2)}R$. The first term in Eq.(5) is linear in perturbation. Our aim is to prove that the first order part of the term $\mathcal{R}_{ab}k^ak^b$ vanishes identically. In order to see that, we note that the terms in $\mathcal{R}_{ab}$ involve squares of curvatures. We are only interested in quantities first order in perturbation over a background stationary spacetime. The background spacetime in our context has no particular physical meaning except as a reference spacetime. Therefore, when we encounter a product of two quantities X and Y, to extract the part linear in perturbation, we will always express such a product as,$XY \approx X^{(B)}Y^{(P)} + X^{(P)}Y^{(B)}$, where $X^{(B)}$ is the value of the quantity X evaluated on the stationary background and $X^{(P)}$ is the perturbed value of X linear in perturbation. Note that, on the stationary background, Raychaudhuri equation demands $R^{(B)}_{ab}k^ak^b = 0$ and since $T^{(B)}_{ab}k^ak^b = 0$, we have $H^{(B)}_{ab}k^ak^b = 0$. Also, to simplify the calculation, we use diffeomorphism freedom to make the null geodesic generators of the event horizon of the perturbed black hole coincide with the null geodesic generators of the background stationary black hole [7]. As a result, the perturbation in the location of the horizon vanishes.

As an illustration of the perturbation scheme, we evaluate a term in $H_{ab}k^ak^b$ as,$2R^{ij}R_{aibj}k^ak^b == 2\left(C_2R^{(P)}_{ab} + R^{(P)ij}\, R^{(B)}_{aibj}\right)\, k^ak^b$ where we have used the results of [9] to express the background Ricci tensor on the horizon as $R^{(B)}_{ab} = C_1 g^{\perp}_{ab} + C_2\gamma_{ab}$. Also, note that $\gamma^{ij}R^{(P)}_{aibj}k^ak^b = R^{(P)}_{ab}k^ak^b$. Implementing similar scheme for other terms, the first order part $\mathcal{R}_1$

of $\left(H_{ab} - 2^{(D-2)}RR_{ab}\right) k^a k^b$ becomes,

$$\mathcal{R}_1 = \left[-\frac{4\,D(D-3)\,R^{(B)}}{(D-1)(D-2)^2} - \frac{16\,(D-3)\,R^{(B)}_{ab}\,k^a l^b}{(D-2)^2} - \frac{8\,C^{(B)}_{abcd}\,k^a l^b l^c k^d}{(D-2)} \right] R^{(P)}_{mn}\,k^m k^n. \qquad (6)$$

The last term in $\mathcal{R}_{ab}k^a k^b$ requires special care. To evaluate this term, we use the standard result of variation of Ricci scalar, neglect a surface term and use the Raychaudhuri equation and the evolution equation for shear [8] keeping terms linear in perturbation to get

$$\frac{d^2}{d\lambda^2}\left({}^{(D-2)}R\right) = 2\,{}^{(D-2)}R^{(B)ab}\,R^{(P)}_{acbd}k^c k^d, \qquad (7)$$

The last line follows after expressing the perturbed Weyl tensor in terms of curvature and Ricci tensors. We rewrite Eq.(7) using Eq.(3) and express $R^{(B)}_{smrn}$ in terms of Weyl and Ricci tensors of the background to arrive at,

$$\frac{d^2}{d\lambda^2}\left({}^{(D-2)}R\right) = \left[\frac{4\,C^{(B)}_{abcd}\,k^a l^b l^c k^d}{(D-2)} - \frac{2\,R^{(B)}\,(D-3)}{(D-1)(D-2)} + \frac{4\,C_2(D-3)}{(D-2)}\right] R^{(P)}_{mn}\,k^m k^n. \qquad (8)$$

Next, using the expression $C_2 = R^{(B)}_{ab}\,\gamma^{ab}/(D-2)$, and comparing with Eq.(6), it is straightforward to prove that the first order term in $\mathcal{R}_{ab}k^a k^b$ vanishes identically and we have the result,

$$\Delta S = 2\pi \int_{\mathcal{H}} \lambda\, T_{ab} k^a k^b \, d\lambda\, \sqrt{\gamma}\, dA + \mathcal{O}(\epsilon^2). \qquad (9)$$

For the background stationary horizon, the Killing vector ξ^a is related with the horizon generators k^a as $\xi^a = \kappa\lambda k^a$, since T_{ab} itself is of $\mathcal{O}(\epsilon)$, we can use this relation in Eq.(9). Then, identifying the Hawking temperature as $T = \kappa/2\pi$, in the leading order of perturbation, we finally obtain,

$$T\Delta S = -\int_{\mathcal{H}} T_{ab}\,\xi^a\, dH^b. \qquad (10)$$

This is the desired form which establishes the physical process version of the first law for EGB gravity. If the matter stress tensor obeys null energy condition, Eq.(9) shows that the Wald entropy for stationary black holes in EGB theory can not decrease in a dynamical process which perturbs the black hole and leads to a new stationary state. Since, it is expected that classical matter obeys null energy condition, we can conclude that as in case of GR, the entropy of stationary black holes in EGB theory can not decrease through any classical process.

[1] J. M. Bardeen, B. Carter, S. W. Hawking, Commun. Math. Phys. **31**, 161-170 (1973).
[2] I. Racz, R. M. Wald, Class. Quant. Grav. **9**, 2643-2656 (1992).
[3] R. M. Wald, Phys. Rev. **D48**, 3427-3431 (1993); V. Iyer and R. M. Wald, Phys. Rev. D **50**, 846 (1994).
[4] S. W. Hawking, Commun. Math. Phys. **25**, 152 (1972).
[5] S. W. Hawking and J. B. Hartle, Commun. Math. Phys. **27**, 283 (1972).
[6] A. Chatterjee and S. Sarkar, Phys. Rev. Lett. **108**, 091301 (2012).
[7] S. Gao and R. M. Wald, Phys. Rev. D **64**, 084020 (2001) [arXiv:gr-qc/0106071].
[8] R. M. Wald, "General Relativity," *Chicago, Usa: Univ. Pr. (1984) 491p*
[9] A. J. M. Medved, D. Martin and M. Visser, Phys. Rev. D **70**, 024009 (2004) [arXiv:gr-qc/0403026].
[10] M. Ortaggio, V. Pravda and A. Pravdova, Class. Quant. Grav. **24**, 1657 (2007) [arXiv:gr-qc/0701150].
[11] E. Gourgoulhon, J. L. Jaramillo, Phys. Rept. **423**, 159-294 (2006). [gr-qc/0503113].
[12] T. Jacobson and R. C. Myers, Phys. Rev. Lett. **70**, 3684 (1993) [arXiv:hep-th/9305016].

Journal of Physics: Conference Series **405** (2012) 012009 doi:10.1088/1742-6596/405/1/012009

Brane inflation: A field theory approach in background supergravity

Sayantan Choudhury[1] **and Supratik Pal**[1,2]
[1]Physics and Applied Mathematics Unit, Indian Statistical Institute, 203 B.T. Road, Kolkata 700 108, India
[2]Bethe Center for Theoretical Physics and Physikalisches Institut der Universität Bonn, Nussallee 12, 53115 Bonn, Germany

E-mail: sayanphysicsisi@gmail.com, supratik@isical.ac.in

Abstract. We propose a model of inflation in the framework of brane cosmology driven by background supergravity. Starting from bulk supergravity we construct the inflaton potential on the brane and employ it to investigate for the consequences to inflationary paradigm. To this end, we derive the expressions for the important parameters in brane inflation, which are somewhat different from their counterparts in standard cosmology, using the one loop radiative corrected potential. We further estimate the observable parameters and find them to fit well with recent observational data. We have studied extensively reheating phenomenology, which explains the thermal history of the universe and leptogenesis through the production of thermal gravitino pertaining to the particle physics phenomenology of the early universe.

1. Introduction

Investigations for the crucial role of Supergravity in explaining cosmological inflation date back to early eighties of the last century [1],[2]. One of the generic features of the inflationary paradigm based on SUGRA is the well-known η-problem, which appears in the F-term inflation due to the fact that the energy scale of F-term inflation is induced by all the couplings via vacuum energy density. Precisely, in the expression of F-term inflationary potential a factor $\exp(K/M_{PL})$ appears, leading to the second slow roll parameter $\eta \gg 1$, thereby violating an essential condition for slow roll inflation. The usual wayout is to impose additional symmetry like Nambu-Goldstone shift symmetry or alternatively to apply Heisenberg symmetry [3] to solve η-problem. Another possible way out to smoothen this problem is fine tuning mechanism via the *radion* fields -which decides the separation between visible and invisible brane or so as to say brane tension in Randall-Sundrum two-brane scenario where the fifth dimension of 5 D bulk is compactified on the orbifold S^1/Z_2 of comoving radius R. This is essential for observationally constraint cosmology on the brane. As we will find in the present article the proposed model of brane inflation matches quite well with latest observational data from WMAP [4] and is expected to fit well with upcoming data from Planck [5]. Next we have studied in detail reheating and leptogenesis, which is different in the context of brane inflation results in novel features. Hence we have established that this has serious implication for the production of the heavy Majorana neutrinos needed for leptogenesis as well as in the gravitino phenomenology through analytical and numerical estimations.

2. Model building from background supergravity

For systematic development of the formalism, let us briefly demonstrate the construction the effective 4D inflationary potential starting from $N=2, D=5$ SUGRA in the bulk which leads to an effective $N=1, D=4$ SUGRA in the brane [3]. Considering S^1/Z_2 orbifold compactification in comoving radius R, the $N=2, D=5$ bulk SUGRA is described by the following action

$$S=\frac{1}{2}\int d^4x\int_{-\pi R}^{+\pi R}dy\sqrt{g_5}\left[M_5^3\left(R_{(5)}-2\Lambda_5\right)+L^{(5)}_{SUGRA}+\sum_{i=1}^{2}\delta(y-y_i)L_{4i}\right]. \quad (1)$$

Here the orbifold points are $y_i=(0,\pi R)$ and $x^m=(x^\alpha,y)$, where y is bounded in closed interval $[-\pi R,+\pi R]$. Written explicitly, the contribution from bulk SUGRA in the action

$$e^{-1}_{(5)}L^{(5)}_{SUGRA}=-\frac{M_5^3R^{(5)}}{2}+\frac{i}{2}\bar{\Psi}_{i\tilde{m}}\Gamma^{\tilde{m}\tilde{n}\tilde{q}}\nabla_{\tilde{n}}\Psi^i_{\tilde{q}}-S_{IJ}F^I_{\tilde{m}\tilde{n}}F^{I\tilde{m}\tilde{n}}-\frac{1}{2}g_{\alpha\beta}(D_{\tilde{m}}\phi^\mu)(D^{\tilde{m}}\phi^\nu)+\text{Fermionic}+\text{Chern}-\text{Simons}, \quad (2)$$

and including the radion fields $(\chi,T,T^\dagger)$ the effective brane SUGRA counterpart turns out to be $\delta(y)L_4=-e_{(5)}\Delta(y)\left[(\partial_\alpha\phi)^\dagger(\partial^\alpha\phi)+i\bar{\chi}\bar{\sigma}^\alpha D_\alpha\chi\right]$. Gauging away the Chern-Simons terms assuming cubic constraints and Z_2 symmetry and appliyng S^1/Z_2 orbifold compactification the effective 4-dimensional action can be expressed as,

$$S=\frac{M^2_{PL}}{2}\int d^4x\sqrt{g_4}\left[R_{(4)}+(\partial_\alpha\phi^\mu)^\dagger(\partial^\alpha\phi_\mu)-QV_F-P\int_{-\pi R}^{+\pi R}dy\frac{4(3e^{2\beta y}+3\lambda^2e^{-2\beta y}-2\lambda)}{R^2(e^{\beta y}+\lambda e^{-\beta y})^5}\right]. \quad (3)$$

where the constants are defined in [3]. Now using the Kähler potential $K=\sum_\alpha\phi^\dagger_\alpha\phi^\alpha$ and the superpotential $W=\sum_{n=0}^{\infty}D_nW_n(\phi^\alpha)$ with $D_0=1$ and Z_2 symmetry the renomalizable effective one-loop F-term potential $N=1, D=4$ SUGRA in the brane turns out to be [3]

$$V(\phi)=\Delta^4\left[1+\left(D_4+K_4\ln\left(\frac{\phi}{M}\right)\right)\left(\frac{\phi}{M}\right)^4\right], \quad (4)$$

where K_4 and D_4 are one loop level constants. For our model energy scale $\Delta\simeq 0.2\times 10^{16}GeV$ for the window $-0.70<D_4<-0.60$.

3. Brane inflation and observational parameters

Here we start with Friedmann equations in brane $H^2=\frac{8\pi V}{3M^2_{PL}}\left[1+\frac{V}{2\lambda}\right]$. Incorporating the potential of our consideration from Eq (4) the new slow roll parameters turn out to be [6]

$$\epsilon_V=\left(\frac{V'}{V}\right)^2\frac{M^2_{PL}(1+\frac{V}{\lambda})}{16\pi(1+\frac{V}{2\lambda})^2},\eta_V=\left(\frac{V''}{V}\right)\frac{M^2_{PL}}{8\pi(1+\frac{V}{2\lambda})},\xi_V=\left(\frac{V'V'''}{V^2}\right)\frac{M^4_{PL}}{(8\pi)^2(1+\frac{V}{2\lambda})^2},\sigma_V=\frac{(V')^2V''''}{V^3}\frac{M^6_{PL}}{(8\pi)^3(1+\frac{V}{2\lambda})^3}, \quad (5)$$

Most significantly the generic SUGRA η-problem is smoothened to some extent in brane cosmology due to the presence of V/λ term in Friedmann equation. The number of e-foldings are defined in brane cosmology for our model as

$$N\simeq\frac{8\pi}{M^2_{PL}}\int_{\phi_f}^{\phi_i}\left(\frac{V}{V'}\right)\left(1+\frac{V}{2\lambda}\right)d\phi\simeq\frac{M^2}{2U}\left[\left(1+\frac{\alpha}{2}\right)\left(\frac{1}{\phi_f^2}-\frac{1}{\phi_i^2}\right)+\frac{D_4}{M^4}(1+\alpha)(\phi_i^2-\phi_f^2)+\frac{\alpha D_4^2}{6M^8}(\phi_i^6-\phi_f^6)\right], \quad (6)$$

where $U=(D_4+4K_4)$. After that the expressions for amplitude of the scalar and tensor perturbation and tensor to scalar ratio [6] are given by

$$\Delta_s^2\simeq\frac{512\pi}{75M^6_{PL}}\left[\frac{V^3}{(V')^2}\left[1+\frac{V}{2\lambda}\right]^3\right]_\star,\Delta_t^2\simeq\left[\frac{\frac{32}{75M^4_{PL}}V\left[1+\frac{V}{2\lambda}\right]}{\sqrt{1+\frac{2V}{\lambda}(1+\frac{V}{2\lambda})}-\frac{2V}{\lambda}(1+\frac{V}{2\lambda})\sinh^{-1}\left[\frac{1}{\sqrt{\frac{2V}{\lambda}(1+\frac{V}{2\lambda})}}\right]}\right]_\star,r=16\frac{\Delta_t^2}{\Delta_s^2}. \quad (7)$$

Here $\star$ represents at the horizon crossing ($k = aH$). Further, the scale dependence of the perturbations, described by the scalar and tensor spectral indices and their running as follows

$$\begin{array}{ll} n_s - 1 = \frac{d\ln\Delta_s^2}{d\ln k}|_\star \simeq (2\eta_V^\star - 6\epsilon_V^\star), & n_t = \frac{d\ln\Delta_t^2}{d\ln k}|_\star \simeq -3\epsilon_V^\star, \\ \alpha_s = \frac{d\ln n_s}{d\ln k}|_\star = (16\eta^\star\epsilon^\star - 18(\epsilon^\star)^2 - 2\xi^\star), & \alpha_t = \frac{d\ln n_t}{d\ln k}|_\star = (6\epsilon^\star\eta^\star - 9(\epsilon^\star)^2), \end{array} \tag{8}$$

where $d(\ln(k)) = Hdt$. Here the consistency conditions $r = 24\epsilon_V = 24\epsilon_V^\star$; $n_t = -3\epsilon_V \simeq -3\epsilon_V^\star = -\frac{r}{8}$ and $\frac{d\sigma}{d(\ln(k))} = (\epsilon\sigma - 2\eta\sigma)$ are valid in brane cosmology.

4. Reheating and Leptogenesis in braneworld

After inflation when reheating epoch starts inflaton decays into different particle constituents. The total inflaton decay width participating in leptogenesis can be written as [7]:

$$\Gamma_{total}(\psi \to l_L\mathcal{H}, \psi \to \bar{l}_L\mathcal{H}) \simeq \frac{C^2}{16\pi m_\phi} + \frac{h^2 m_\psi}{4\pi} \sim \frac{1}{(2\pi)^3}\left(\frac{\Delta^6}{M^5}\right) = 3H(T^{brh}) = \sqrt{\frac{3\rho(t_{reh})}{M^2}\left[1 + \frac{\rho(t_{reh})}{2\lambda}\right]}. \tag{9}$$

Now applying statistical thermodynamics in braneworld the expression for transition temperature($T_{c\gamma}$) and the reheating temperature(T^{brh}) turns out to be

$$T_{c\gamma} = \sqrt[4]{\left\{C_\gamma \frac{\pi^2(K_4+4D_4)^2\Delta_s^2\phi_\star^6}{\pi^2 N_\gamma^\star \alpha^4 M^2}\right\}}, \qquad T_\gamma^{brh} = \sqrt[4]{\left\{\frac{W_\gamma(K_4+4D_4)\Delta_s\phi_\star^3\Gamma_{total}}{\pi N_\gamma^\star \alpha^2}\right\}} \qquad \forall\gamma \tag{10}$$

where $C_\gamma = \left(36000, \frac{288000}{7}, 19200\right)$, $W_\gamma = \left(600, \frac{4800}{7}, 320\right)$, the species index $\gamma = 1(B \Rightarrow Boson), 2(F \Rightarrow Fermion), 3(M \Rightarrow Mixture)$ and $N^\star = N_B^\star + \frac{7}{8}N_F^\star$.

To study the impact of ϕ^4 in leptogenesis we will start with the Boltzmann equation [8]

$$\dot{\rho_r} + 4H\rho_r = \Gamma_\phi\rho_\phi, \tag{11}$$

where in braneworld high energy limit ($\rho \gg \lambda$) $H^2 = \frac{4\pi}{3\lambda M_{PL}^2}(\rho_r + \rho_\phi)^2$.Now extremization of the solution of the equation(11) gives extremum temperature during reheating $T_{ex}^{bh} = \sqrt[4]{\left\{\frac{45\Gamma_\phi M_5^3}{8N^\star\pi^3}\right\}}$.To use this idea in gravitino phenomenology we will start with gravitino Boltzmann equation

$$\frac{dn_{\tilde{G}}}{dt} + 3Hn_{\tilde{G}} = \langle\Sigma_{total}|v|\rangle n^2 - \frac{m_{\frac{3}{2}} n_{\tilde{G}}}{\langle E_{\frac{3}{2}}\rangle\tau_{\frac{3}{2}}}. \tag{12}$$

Henve solving equation(12) using $Y_{\tilde{G}} = \frac{n_{\tilde{G}}}{s}$ gravitnino abundance at the end of reheating can be expressed as[7]:

$$Y_{\tilde{G}}^{brh}(T_f) = \left(\frac{45\zeta^2(3)\tilde{\alpha}\sqrt{3\lambda}}{2\pi^7 M\Delta^2 N^\star}\right)\left[\left(\frac{60\sqrt{\lambda}}{\pi N^\star T_f}\right)\left(1 - \frac{T_f}{T^{brh}}\right) + \left(\frac{\pi (T_{ex}^{bh})^4}{32\Delta^2 T^{brh}}\right)\left(32\left(\frac{T^{brh}}{T_{ex}^{bh}}\right)^4 - 1\right)^{\frac{1}{4}} \times \left\{\frac{1}{\Gamma(\frac{3}{4})} + \frac{1}{\Gamma(-\frac{3}{4})} - \frac{2}{\Gamma(-\frac{1}{4})}\right\}\right]. \tag{13}$$

which is plotted with respect to temperature in figure(1(b)).

5. Numerical estimations and data analysis

For a typical value of $C_4 \simeq D_4 = -0.7$, $56 < N < 70$ and decay width $\Gamma_\phi \simeq 2.9 \times 10^{-3}GeV$, mass of the inflaton $m_\phi \simeq 10^{13}GeV$ we have: $\Delta_s^2 \sim (1.440 - 3.126) \times 10^{-9}$, $\Delta_s^2 \sim 10^{-14}$, $n_s \sim (0.936 - 0.951)$, $n_t \sim 10^{-5}$, $r \sim (2.176 - 4.723) \times 10^{-5}$, $\alpha_s \sim -10^{-3}$, $\alpha_t \sim 10^{-6}$,

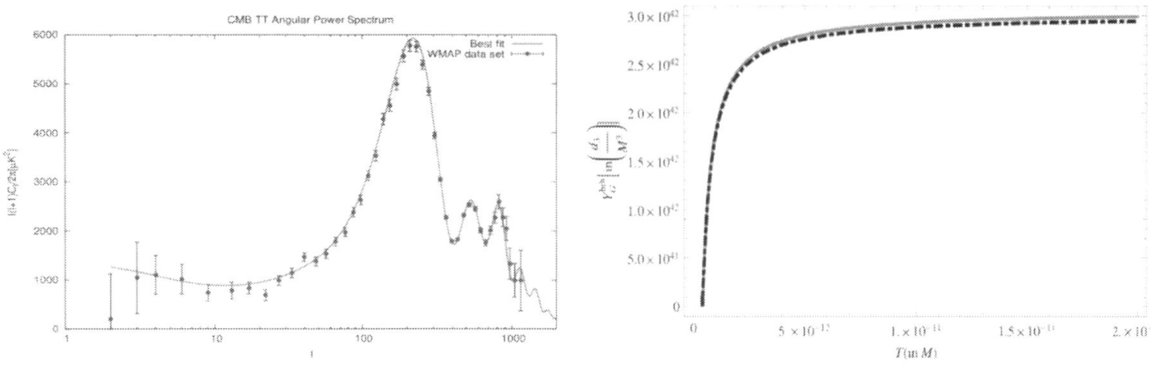

Figure 1. Variation of (a) CMB angular power spectrum C_l^{TT} with the multipoles l for scalar modes, (b)total gravitino abundance vs temperature

for boson $T_B^{brh} \simeq 7.6 \times 10^{10}GeV$, $T_{cB} \simeq 3.2 \times 10^{14}GeV$ for fermion $T_F^{brh} \simeq 7.8 \times 10^{10}GeV$, $T_{cF} \simeq 3.3 \times 10^{14}GeV$ and for mixture of species $T_M^{brh} \simeq 6.5 \times 10^{10}GeV$, $T_{cM} \simeq 2.8 \times 10^{14}GeV$, $T_{ex}^{bh} \simeq 7.0 \times 10^{10}GeV$, final temperature $T_f \simeq 10^6GeV$, at the end of reheating $Y_{\tilde{G}}^{b-rad}(T_f) \simeq 2.1 \times 10^{-13}GeV^{-3}d_3$, where $d_3 = 6.594\tilde{\alpha} \times 10^{11}GeV^3$ and $\tilde{\alpha}$ is a dimensionless constant in MSSM. Now using CAMB [9] with the pivot scale $k_0 = 0.002$ Mpc^{-1}, $H_0 \sim 71.0km/sec/MPc$, $\tau_{Reion} \sim 0.09$, $\Omega_b h^2 \sim 0.022$, $\Omega_c h^2 \sim 0.111$ and $T_{CMB} \sim 2.725K$ we get $t_0 \sim 13.707Gyr$, $z_{Reion} \sim 10.704$, $\Omega_m \sim 0.267$, $\Omega_\Lambda \sim 0.732$, $\Omega_k \sim 0$, $\eta_{Rec} \sim 285.10$ and $\eta_0 \sim 14345.1$.

6. Summury

In this article we have proposed a model of brane inflation derived from $N = 2, D = 5$ supergravity in the bulk leads to an effective $N = 1, D = 4$ supergravity in the brane including one-loop radiative corrections. Then the model has been employed in estimating the observable parameters, both analytically and numerically, leading to more precise estimation of the quantities using the publicly available code CAMB [9], which reveals consistency of our model with latest observations. Hence we have studied reheating and leptogenesis in brane by analyzing the reheating temperature, followed by analytical and numerical estimation of different phenomenological parameters showing deviations from GR results.

Acknowledgments

SC thanks Council of Scientific and Industrial Research, India for financial support through Junior Research Fellowship (Grant No. 09/093(0132)/2010).

References

[1] H. P. Nilles, *Phys. Rept.* (1984) **110**.
[2] D. H. Lyth and A. Riotto, *Phys. Rept.* (1999) 314 and references therein.
[3] S. Choudhury and S. Pal, *Phys. Rev. D* 85 (2012)**043529**.
[4] For uptodate results on WMAP, see *http://lambda.gsfc.nasa.gov/product/map/current*.
[5] Planck collaboration, *http://www.rssd.esa.int/index.php?project=Planck*.
[6] R. Maartens and K. Koyama, *Living Rev. Relativity* 13, (2010) **5**.
[7] S. Choudhury and S. Pal, *Nucl. Phys. B* 857 (2012) **85**.
[8] E. Kolb and M. Turner, The Early Universe,*Addison Wesley* (1990).
[9] CAMB, Online link: *http://camb.info/*.

Conformally invariant Inert Higgs doublet model : an unified model for Inflation and Dark matter

Moumita Das[1] **and Subhendra Mohanty**[2]
Physical Research Laboratory, Ahmedabad 380009, India
E-mail: [1] moumita@prl.res.in, [2]mohanty@prl.res.in

Abstract. Motivation of our present study is the searching for an unified model which can describe both the inflation as well as dark matter. From particle physics point of view, Higgs can be the most interesting candidate for the scalar field inflation. Conformal coupling of the inflaton with the gravity can generate the density perturbation and we use this idea in a realistic inert Higgs doublet model. We study the loop corrections of this conformally coupled system and in present era there is electroweak symmetry breaking to provide the mass of the particles. Study of the mass spectrum in present era reveals the scalar dark matter with mass 33.7 GeV and lightest Higgs at 125.6 GeV.

1. Introduction

Scalar field inflation is most successful model of inflation. Higgs can be the most favorable candidate for this kind of model. Introduction of Standard Model Higgs as inflaton was done by F.Bezrukov and M. Shaposhnikov [1]. Generating inflation from the potential of the form $\lambda\phi^4$ demands that λ should be $\mathcal{O}(10^{-12})$. For standard model Higgs, λ is approximately ~ 1 and the Higgs can not be used as inflaton. They solved this issue by considering the large coupling ($\xi \sim 10^4$) between Ricci scalar and the Higgs [1]. The problem of unitarity of graviton-scalar scattering [2] causes due to the presence of large coupling ξ. However there are some attempts [3] present in the literature to solve this issue.

Rubakov and collaborators have shown different approach for the generation of scale invariant density perturbations [4]. The idea was that a conformally coupled field rolling down a quartic potential can generate scale invariant density perturbation. These perturbations can become superhorizon in an inflationary era or in a ekpyrotic scenario. We implement this idea in a realistic inert Higgs doublet model. The study of electroweak symmetry generate the mass spectrum in present era. The requirement of scale invariance at high energy scale and electroweak symmetry breaking at low energies fixes the coupling constants of the theory. We find that there is a set of benchmark parameters of this model where the mass of the Higgs boson is $m_h = 125.6$ GeV and the mass of the dark matter (which is the lightest neutral component of the inert doublet) $m_{A_0} = 33.7$ GeV, which can be tested at the LHC and in direct detection experiments. The Higgs mass is consistent with the 2.8σ signal of a 125 GeV higgs decay into $\gamma\gamma$ and $4l$ observed by the ATLAS [5] and CMS [6] collaborations.

2. Inert Doublet Model and Coleman-Weinberg loop correction

Inert Doublet Model (IDM) respect the Z_2 symmetry, under which all Standard model particles including the SM Higgs H_1 are even and an extra scalar doublet H_2 is odd. Due to Z_2 symmetry, the cubic term and Yukawa term for H_2 doublet are forbidden and its neutral component can be a candidate for dark matter. The most general renormalizable potential will be,

$$V_{tree} = V_c + \mu_1^2|H_1|^2 + \mu_2^2|H_2|^2 + \lambda_1|H_1|^4 + \lambda_2|H_2|^4 + \lambda_3|H_1|^2|H_2|^2 + \lambda_4|H_1^\dagger H_2|^2 + \frac{\lambda_5}{2}\left[\left(H_1^\dagger H_2\right)^2 + h.c.\right] \tag{1}$$

where the two Higgs doublets H_1 and H_2 are, $H_1 = \begin{pmatrix} h^+ \\ \frac{h+iG_0}{\sqrt{2}} \end{pmatrix}$ and $\mathrm{H}_2 = \begin{pmatrix} H^+ \\ \frac{H_0+iA_0}{\sqrt{2}} \end{pmatrix}$

We consider the conformal case where $\mu_1 = \mu_2 = 0$ and the constant potential, $V_c = 4\times 10^7\,\mathrm{GeV}^4$ such that the minimum of the total potential becomes zero at present era. We derive the one-loop correction to the potential (1) following Coleman-Weinberg formalism [7] and the effective potential becomes,

$$V_{eff.}(H_0, h, \mu) = V_{tree} + \frac{1}{64\pi^2}\sum_i n_i m_i^4 \ln(\frac{m_i^2}{\mu^2} + 1) \tag{2}$$

where n_i is the degree of freedom and m_i are tree level masses. In this case we have also considered the loop corrections from the top quark and gauge bosons. However the gauge boson loop has negligible effect on the results.

To get the correct electro-weak symmetry breaking in the present era and the scale invariant density perturbation in the early era, we have chosen a set of $\lambda_i = \{-0.14, -11, 2.8, -1.52, -1.52\}$ in present epoch. Now we have to study the running of couplings λ_i, where $\{i = 1\,\mathrm{to}\,5\}$ using the one-loop renormalization group equation for the inert doublet model [8] and we find the $\lambda_i = \{0.49, -0.5, 2.1, 0.84, -2.2\}$ in the early era, $\mu \simeq 10^5$ GeV.

3. Generation of the scale invariant density perturbation

We now turn to the question of the generation of density perturbations in the early era when V_{eff} (2) simplifies to the form $V_{inf} \sim V_c + \frac{\lambda_2}{4} H_0^4$ shown in Fig. (1), where $V_c = 4\times 10^7\,\mathrm{GeV}^4$ and $\lambda_2 = -0.5$. We take the inert Higgs doublet to be conformally coupled to gravity and the

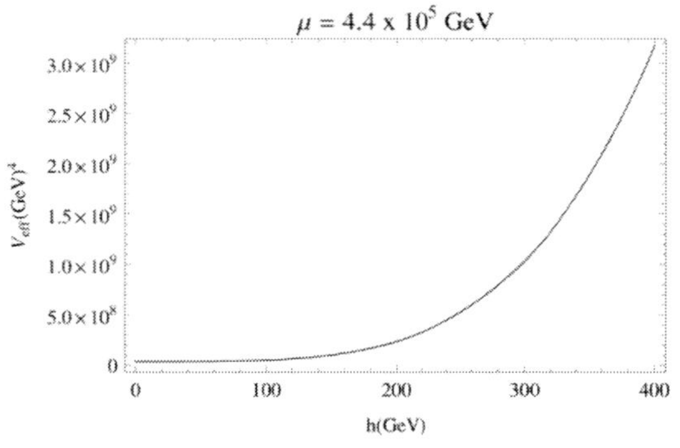

(a)

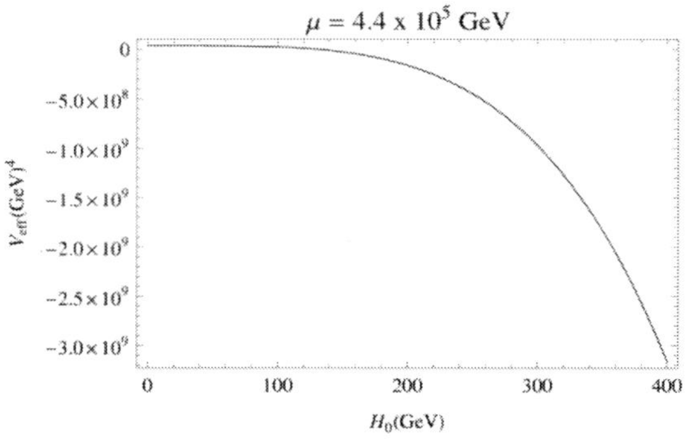

(b)

Figure 1. Effective potential in the early universe

action for this field can be written as,

$$S = \int d^4x\sqrt{-g}\left[g^{\mu\nu}\partial_\mu H_2^*\partial_\nu H_2 - \frac{R}{6}H_2^*H_2 - V_{inf}\right] \tag{3}$$

where H_2 contains the neutral part of the inert doublet i.e. $H_2 = (H_0 + i\,A_0)/\sqrt{2}$ and R is the scalar curvature, which conformally coupled with the field H_2. Defining $H_2 = \chi_{H_2}/a$, we can simplify the equation of the field H_2 as

$$\chi''_{H_2} + k^2\chi_{H_2} + a^3\frac{\partial V_{inf}}{\partial H_2} = 0 \tag{4}$$

The conserved current will be of the form $\frac{d}{d\eta}(\rho^2\theta') = 0$, after expressing, $\chi_{H_2} = \rho\exp(i\,\theta)$. Hence, the field rolls along the radial direction while the phase θ remains constant with the increase of ρ. Without loss of generality we can choose the fixed phase such that the field H_2 has only real and neutral component χ_{H_0}. The perturbations of H_2 will be along the imaginary axis and we can denote the full H_2 with the perturbations as, $\chi_{H_2} = \chi_{H_0} + i\,\delta\chi_{A_0}$. From Eq (4), the equation of motion of χ_{H_0} will be,

$$\chi''_{H_0} + k^2\chi_{H_0} - \frac{\lambda_2}{2}\chi^3_{H_0} = 0 \tag{5}$$

Considering $k \ll 1/\eta$ at late time, the solution will be, $\chi_{H_0} \approx 1/\left(\sqrt{-\lambda_2}(\eta_* - \eta)\right)$, where $\sqrt{-\lambda_2}$ is a real quantity as λ_2 is negative and η_* is a constant of integration. At the end of inflation, when $\mu << 10^5$, the shape of the potential changes to Fig. (2) and H_0 starts rolling back to zero.

The equation of motion of the perturbation, $\delta\chi_{A_0}$ is given by,

$$\delta\chi''_{A_0} + k^2\delta\chi_{A_0} + \frac{\lambda_2}{2}\chi^2_{H_0}\delta\chi_{A_0} = 0 \tag{6}$$

and at later times, when $(k(\eta_* - \eta) \ll 1)$, solution will be $\delta\chi_{A_0} \sim 1/\left(k^{3/2}(\eta_* - \eta)\right)$. Hence, the super-horizon perturbations of the phase can be defined as $\delta\theta \equiv \delta\chi_{A_0}/\chi_{H_0}$. The power spectrum of $\delta\theta$ is scale invariant as follows,

$$\mathcal{P}_{\delta\theta} = \frac{k^3}{2\pi^2}|\delta\theta|^2 = \frac{-\lambda_2}{2\pi^2} \quad \text{where} \quad \delta\theta = \frac{\sqrt{-\lambda_2}}{\mathrm{k}^{3/2}} \tag{7}$$

If one considers the k dependence of the equation of motion of H_0 as discussed in [9], there will be a deviation from the scale free power spectrum (7) and it will give rise to a non-zero spectral index, $n_s - 1 = \frac{3\lambda_2}{4\pi^2} = -0.04$, which is consistent with the WMAP observation of $n_s = 0.967 \pm 0.014$ [10]. The perturbations of the phase $\delta\theta = \delta A_0/H_0$ can be converted to adiabatic perturbation by the decay of the A_0 field into standard model fields as in the curvaton mechanism [11]. The amplitude of adiabatic perturbation is related to the phase perturbation as

$$P_\zeta = r^2\frac{P_{\delta\theta}}{\theta_c^2} = r^2\frac{-\lambda_2}{2\pi^2\theta_c^2} \tag{8}$$

where r is the ratio of the energy density in the A_0 field oscillations to the total energy density. Taking the unperturbed phase to be $\theta_c \sim \pi/2$, and with $\lambda_2 = -0.5$, we see that $r = 2\times10^{-4}$ is needed to give the required $P_\zeta = 10^{-10}$.

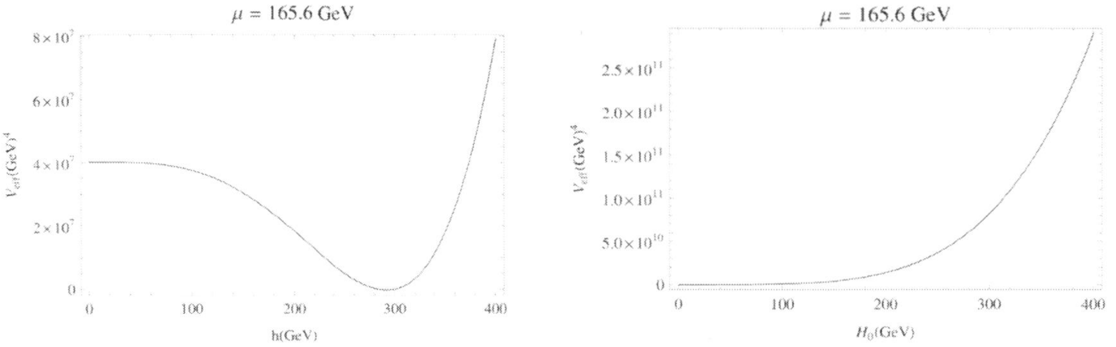

(a) (b)

Figure 2. Effective potential in the present universe

4. Scalar mass spectrum

As the field H_2 has a zero vev in the present universe, the lightest neutral components of H_2 will be stable and can be the candidates for dark matter. We study the masses of the fields in present universe from the effective potential taking $< H_1 >= 246\,\text{GeV}$ and $< H_2 >= 0\,\text{GeV}$. We see that the field A_0 can be a candidate for light dark matter with mass $M_{A^0} = 33.7$ and the predicted Higgs mass is $M_h = 125.6$ GeV which matches with the current observation of Atlas [5] and CMS [6] collaborations. The mass of the other scalars H^0 and $H^{\pm}$ are 273.6 and 433.5 respectively.

5. Conclusions

The inert Higgs doublet model gives a natural extension of the standard model and can be used for explaining the electroweak symmetry breaking by loop corrections [12] starting from a scale invariant tree level potential. We find that the the quartic coupling of the inert doublet, $\lambda_2 = -0.5$ at $\mu = 10^5$ GeV, predicts correct spectral index of the power spectrum of the perturbations. The amplitude of the power spectrum P_ζ also can be tuned to be consistent with the observations by choosing a suitable curvaton mechanism. Finally our model works for the chosen parameters λ_i, to give the Higgs mass at $m_h = 125.6$GeV matches with the observations of ATLAS [5] and CMC[6]. The same set of λ_i also offers a candidate for dark matter with a mass $M_{A^0} = 33.7$ GeV which may possibly be observed in direct detection experiment.

References

[1] Bezrukov F L and Shaposhnikov M, 2008, *Phys. Lett.* **B659**, 703-706 . (arXiv : 0710.3755).
[2] Hertzberg M P, 2010, *JHEP* **1011**, 023 (arXiv : 1002.2995).
[3] Germani C and Kehagias A, 2010, *Phys. Rev. Lett.* **105**, 011302 (arXiv : 1003.2635).
[4] Rubakov V A, 2009, *JCAP* **0909**, 030 (arXiv : 0906.3693).
[5] Aad G *et al.* [ATLAS],2012, *Phys. Lett. B* **710**, 49; (arXiv : 1202.1408); Aad G *et al.* [ATLAS],(arXiv:1207.7214).
[6] Chatrchyan S *et al.* [CMS],2012, *Phys. Lett. B* **710**, 26; (arXiv : 1202.1488); Chatrchyan S *et al.* [CMS],Phys. Lett. B (arXiv:1207.7235).
[7] Coleman S R and Weinberg E J, 1973, *Phys. Rev. D* **7**, 1888.
[8] G. C. Branco, P. M. Ferreira, L. Lavoura, M. N. Rebelo, M. Sher and J. P. Silva, 2012, Phys. Rept. **516** 1
[9] M. Libanov, S. Ramazanov and V. Rubakov, 2011, *JCAP* **1106**, 010 (arXiv : 1102.1390).
[10] Komatsu E *et al*, 2011, *Astrophys. J. Suppl.* **192** 18 (arXiv : 1001.4538).
[11] Lyth D H and Wands D, 2002, *Phys. Lett. B* **524**, 5 (arXiv : hep-ph/0110002).
[12] Hambye T and Tytgat M H G, 2008, *Phys. Lett. B* **659**, 651 (arXiv : 0707.0633).

Sterile neutrino, hidden dark matter and their cosmological signatures

Subinoy Das

Institut für Theoretische Teilchenphysik und Kosmologie
RWTH Aachen, D-52056 Aachen, Germany

E-mail: subinoy@physik.rwth-aachen.de

Abstract. Though thermal dark matter has been the central idea behind the dark matter candidates, it is highly possible that dark matter of the universe is non-thermal in origin or it might be in thermal contact with some hidden or dark sector but not with standard model. Here we explore the cosmological bounds as well as the signatures on two types of non-thermal dark matter candidates. First we discuss a hidden dark matter with almost no interaction (or very feeble) with standard model particles so that it is not in thermal contact with visible sector but we assume it is thermalized with in a hidden sector due to some interaction. While encompassing the standard cold WIMP scenario, we do not require the freeze-out process to be non-relativistic. Rather, freeze-out may also occur when dark matter particles are semi-relativistic or relativistic. Especially we focus on the warm dark matter scenario in this set up and find the constraints on the warm dark matter mass, cross-section and hidden to visible sector temperature ratio which accounts for the observed dark-matter density, satisfies the Tremaine-Gunn bound on dark-matter phase space density and has a free-streaming length consistent with cosmological constraints on the matter power spectrum. Our method can also be applied to keV sterile neutrino dark matter which is not thermalized with standard model but is thermalized with in a dark sector. The second part of this proceeding focuses on an exotic dark matter candidate which arises from the existence of eV mass sterile neutrino through a late phase transition. Due to existence of a strong scalar force the light sterile states get trapped into stable degenerate micro nuggets. We find that its signature in matter power spectra is close to a warm dark matter candidate.

1. Introduction

Here we briefly discuss a unified treatment [2] of the freeze-out of thermal relics in hidden sectors for arbitrary cross section σ, only requiring that constraints from cosmology are satisfied, and find viable dark matter that freezes out when it is relativistic, semi-relativistic, or non-relativistic. Our result applies to sterile neutrino warm dark matter which might or might not be in thermal equilibrium with visible sector. Especially, if the dark matter sterile neutrinos are not in thermal equilibrium with visible sector but somehow has enough interactions in the dark sector to thermalize it, this lower bound of 1.5 keV is pretty generic and applies in such situation.

In the last part of this proceeding we focus on recent hints by MiniBooNe and reactor experiment which along with LSND may point out to the existence of eV scale sterile neutrino. We discuss an exotic scenario where being triggered by a scalar field such eV sterile states which was highly relativistic deep in radiation dominated era went through a phase transition few e-foldings before matter radiation equality and clumped into micro-nuggets. These nuggets composed of many sterile states can be very heavy –mass ranging from keV to TeV. But the signature of the this dark matter are always close to that of warm dark matter particle as far as matter power spectra is concerned. This is due the fact that light sterile neutrino behaves like hot dark matter before the phase transition and then transitions into cold dark matter nuggets. This means we always get suppression in power in smaller scales in the matter power spectra and it mimics CDM at larger scales.

2. Constraints on hidden warm dark matter

For a temperature-dependent dark matter cross section $\sigma f(x)$, a general (warm, cold or hot) freeze-out is described by [2]

$$\frac{dy}{dx} = -\frac{1}{x^2}\left[y^2 - y_0^2(x;\xi)\right] + y\,\frac{d\ln f(x)}{dx}\,. \tag{1}$$

where $y(x;\xi)$ is the scaled yield of dark matter and $f(x) = \left(\frac{3}{x^2} + \frac{\frac{5}{4}+x}{1+x}\right)$ with $x = m/T$. Numerically we solve this equation for different regimes of freeze-out and find the hidden dark matter relic density which depends on dark matter mass (m_χ), cross-section ($\sigma(x)$) and hidden to visible sector temperature ratio (ξ).

Other than the correct dark matter relic density, we now discuss two more cosmological constraints which applies to hidden or visible sector dark matter candidate. Generally, hidden-sector dark matter does not need to freeze-out when non-relativistic (cold) in order to match observations. The dark-matter particles may have significant thermal momentum at freeze-out (it can be warm or hot from the hidden-sector perspective), and this can lead to a significant free-streaming length that can suppress the matter power spectrum on scales as large as those corresponding to galactic length scales. While the freeze-out process is thermal in the hidden sector, since the hidden sector is at a different temperature than the visible sector, it mimics a non-thermal freeze-out process.

$$\lambda^{FSH} = \frac{1}{\sqrt{\Omega_r}H_0}\left[\int_{a_f}^{a_\lambda} da + \int_{a_\lambda}^{a_e} \frac{a_{nr}\,da}{\sqrt{a^2 + \frac{a^3}{a_e}}}\right]\,. \tag{2}$$

where a_e is the scale factor of the Universe at matter-radiation equality. Within our approximation, prior to a_{nr} we take $< v >= 1$ while for $a > a_{nr}$ we take $< v > \propto a^{-1}$, and $a_\lambda = \max[a_f, a_{nr}]$. Here, a_{nr} corresponds to a scale factor when dark matter particles becomes nonrelativistic ($T^h_{nr} \simeq m_\chi/3.15$) Here we use here the bound $\lambda^{FSH} \leq 230$ kpc to constrain hidden DM parameter space.

A robust and model independent lower bound on the mass of dark matter particles is obtained by bounding the phase-space density evolution of small galaxies like the dwarf spheroidal satellites (dSphs) of the Milky Way. For example, if the dark matter is fermion it gets a stringent bound, independent of cosmological evolution, as Pauli blocking enforces a densest packing of the dark matter phase space distribution. Since the microscopic phase space density (PSD) is exactly conserved for collision less and dissipation less particles by Liouville's theorem the coarse-grained (averaged) PSD can not be an increasing function of time. This means that the maximum coarse-grained PSD in a galaxy today, with core radius r_c and velocity dispersion

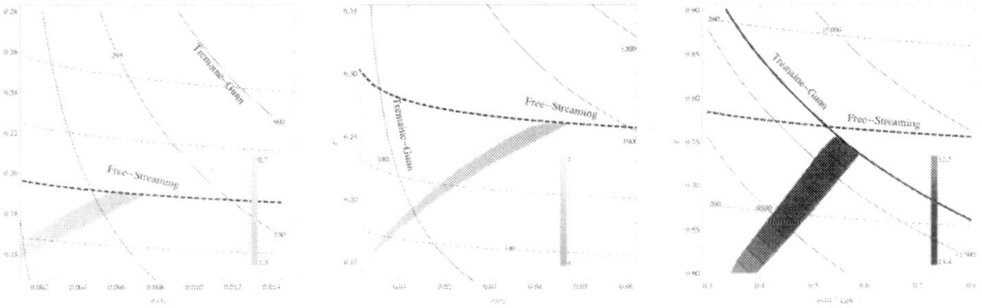

Figure 1. Representative constraints in $(\sigma/\sigma_0, \xi)$ plane for $m_\chi = 1900$ eV (left panel), 3 keV (middle panel), and 12.5 keV (right panel). Here, $\sigma_0 = 10^{-9}$ GeV^{-2}. The horizontal dashed lines are contours of free-streaming length while the bold one corresponds to $\lambda^{FSH} \simeq 230$ kpc. The solid lines are contours of minimum Tremaine-Gunn mass, $m_{min}^{TG}(\xi,\sigma)$. We see that the free-streaming bound is more stringent for such low dark matter mass. The colored region corresponds to the allowed values of dark matter density $(0.1 \leq \Omega_{DM}h^2 \leq 0.114)$ and that also obey the Free-streaming and Tremaine-Gunn bounds. The corresponding values of hidden-sector $x_f^h \equiv m/T_f^h$ are shown in the sidebar to illustrate the nature of decoupling (i.e., whether relativistic, nonrelativistic or semirelativistic). We see that for low mass x_f^h is smaller and decoupling tends to be relativistic, while for higher mass x_f^h is larger with values corresponding to either semirelativistic (warm) or nonrelativistic freeze out.

σ_V, must not exceed the maximum value of the microscopic PSD (the Fermi-Dirac for a fermion) over cosmic history $f_{max}^{cg}(t) \leq f_{max}^{fg}$. This inequality translates to a bound on the mass of

$$m \geq m_{min} = \left(\frac{9\,h^3}{(2\pi)^{5/2} d_\chi\, G_N\, \sigma_V\, r_c^2} \frac{1}{f(q, T_f^h)|_{max}} \right)^{1/4} \qquad (3)$$

In fig. 1 we show the free-streaming constraint, Phase-space bound as well as the bound from correct dark matter relic density on hidden sector dark matter candidates. As expected the the bound becomes stronger for the hidden warm dark matter.

3. Warm dark matter through phase transition from light sterile states

Data from MiniBooNE experiment [3] may indicate the existence of two light sterile neutrino states (eV mass) and it is possible that these light states have non-trivial interactions with, for example, a scalar. Interestingly, recent analysis of WMAP 7 data can not only accommodate but might prefer such extra radiation degrees of freedom like light sterile neutrino states in the early universe.

Motivated by this possibility, here in this paper we propose a new scenario where such light fermions, at some point in the radiation dominated era (RDE) went through a phase transition and clumped into small dark matter nuggets in presence of a scalar-mediated fifth force. Due to its interaction with a scalar, the fermion mass is chameleon in nature and depends on the scalar vacuum expectation value *vev*. So, the dynamics of ϕ is controlled by an effective potential instead of just $V(\phi)$ and the scalar field adiabatically tracks the minima of V_{eff}. As the background fermion density dilutes due to expansion of the universe, the minimum of V_{eff} is time dependent and so is the fermion mass. In our case, the attractive scalar force starts to dominate over the free-streaming when the fermion becomes heavy enough before MRE and results in dark matter micro nuggets[4]. As a result, the fermions within each scalar Compton

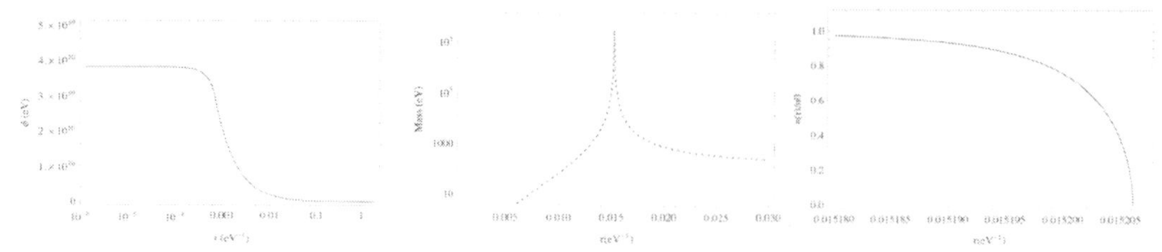

Figure 2. Numerical results for example A. The top panel shows the radial profile of the scalar field (where $eV^{-1} \equiv 2 \times 10^{-5}$cm). The middle panel shows fermion mass as function of distance r, from the nugget center where radius of nugget is 10^{-7} cm. The bottom panel shows the fermion number density $\frac{n}{n_0}$ (n_0 being the density at the center) as a function of distance. The radius is determined when number density drops to zero.

volume collapse into one nugget until the fermi pressure intervenes and balances the attractive force. The scalar field obtains a static profile as the nugget forms. We find that the scalar field inside the nugget is displaced in such a way that the fermion mass inside the nugget is much smaller than outside, ensuring the stability of the nuggets. The radius of the nugget is also determined by the scalar profile which in our case can be as small as micrometer ($R_{nug} \leq \mu$m). We take a lagrangian of the form

$$\mathcal{L} \supset m_D \psi_1 \psi_2 + f(\Phi)\psi_2\psi_2 + V(\Phi) + \text{h.c.}\,, \tag{4}$$

Here ψ_1 is the light fermion field, ψ_2 is heavier one and both of them are written as two component left chiral spinor field while m_D is the Dirac mass term and $V(\phi)$ being the scalar potential. If ψ_2 is considerably heavier, we can integrate it out from the low energy effective theory obtaining $m_\psi = \frac{m_D^2}{f(\phi)}$.

So a rough estimate of the dark matter density at the time of phase transition is given by $\rho_{z_F}^{DM} = \frac{M_{nug}}{\frac{4}{3}\pi(m_\phi^{-1})^3}$ which basically assumes that all the sterile fermions in one scalar Compton volume get trapped into one nugget. Here z_F denotes the redshift of nugget formation. Evolving this density to the present epoch we get $\rho_{DM}^0 = \rho_F^{DM}(1+z_F)^{-3}$ where M_{nug} is the nugget mass which can be found by solving the scalar profile.

As shown in details [5], the nugget profile is obtained by solving two equations – Klein-Gordon equation for the scalar field and force balance equation between the scalar attraction and fermi pressure

$$\phi'' + \frac{2}{r}\phi' = \frac{dV(\phi)}{d\phi} - \frac{dln[m(\phi)]}{d\phi}T^\mu_\mu$$
$$\frac{dp}{d\phi} = \frac{d[ln(m(\phi))]}{d\phi}(3p - \rho)$$

where $m(\phi)$,p,ρ is the scalar field mass, pressure and energy density of the dark matter fermion. Using Thomas-Fermi approximation, one can numerically solve the profile for the dark matter nugget [5] which has been shown in Fig.2.

4. References

[1] J. L. Feng and J. Kumar, Phys. Rev. Lett. **101**, 231301 (2008) [arXiv:0803.4196 [hep-ph]];
[2] S. Das and K. Sigurdson, Phys. Rev. D **85**, 063510 (2012) [arXiv:1012.4458 [astro-ph.CO]].
[3] J. Kopp, M. Maltoni and T. Schwetz, arXiv:1103.4570.
[4] N. Afshordi, M. Zaldarriaga and K. Kohri, Phys. Rev. D **72**, 065024 (2005) [arXiv:astro-ph/0506663].
[5] S. Das and K. Sigurdson, "Degenerate Dark Matter Micronuggets from Ultralight Trapped Fermions", Manuscript in preperation.

International Conference on Modern Perspectives of Cosmology and Gravitation (COSGRAV12) IOP Publishing
Journal of Physics: Conference Series **405** (2012) 012012 doi:10.1088/1742-6596/405/1/012012

Generalized particle dynamics in anti de Sitter spaces: mimicking a phantom dark energy

Sudipta Das[1], Subir Ghosh[1], Jan-Willem van Holten[2], Supratik Pal[1,3]

[1]Physics and Applied Mathematics Unit, Indian Statistical Institute, 203 B.T. Road, Kolkata 700 108, India

[2]NIKHEF, PO Box 41882, 1009 DB Amsterdam, Netherlands

[3]Bethe Center for Theoretical Physics and Physikalisches Institut der Universität Bonn, Nussallee 12, 53115 Bonn, Germany

E-mail: sudipta.das_r@isical.ac.in, sghosh@isical.ac.in, t32@nikhef.nl, supratik@isical.ac.in

Abstract.
We consider a generalized particle dynamics in brane world formalism in an (4+1)-asymptotically anti de Sitter background. This generalized particle dynamics model induces an effective (4+1)-dimensional spacetime and (3+1)-dimensional physical universe is then embedded in this effective spacetime. The embedding framework induces an effective Dark Energy equation of state which is dynamical in nature and exhibits a phantom like behavior. We then corroborate our results with present day observed cosmological parameters.

1. Introduction

Observational evidence of the present-day acceleration of the universe [1], and the subsequent precise measurements of observable parameters [2] indicates that the entity Dark Energy (DE) is responsible for this recent acceleration. The DE density-fraction of total cosmic density ($\Omega_{\rm DE}$) is determined conclusively from several independent probes ($\Omega_{\rm DE} = 0.734 \pm 0.029$ at 95% C.L. from latest WMAP7 data [2]). On the other hand, the analysis of the most reliable SNIa Gold dataset show strong indication that the DE effective Equation of State (EOS) $w_{\rm DE} < -1$ (the lower bound being $-1.12 < w_{\rm DE}$ from WMAP7 data [2]), leading to the conclusion that models with phantom divider crossing are preferred over ΛCDM (or quintessential candidates) at 2σ level. This clearly weakens the claims of cosmological constant Λ or dynamical scalar field models as viable DE candidates. One can resurrect the scalar field models only at the cost of phantom fields with a negative kinetic term but they bring in severe instability problems. In this perspective, it seems reasonable to explore modified gravity theories [3] which do not suffer from any such major drawbacks. It is well-known [3, 4] that by embedding techniques one can relate cosmological surface dynamics (Friedmann equations) in lower (3+1)-dimensions with particle motion in a higher (4+1)-dimensional black-hole like space-time. Our introduction of generalized particle dynamics [5] in such space-times can give rise to an effective negative pressure resulting in cosmic acceleration and the subsequent crossover of the phantom divide. We further establish our model by an analysis of the equation of state and a determination of the relevant parameters describing the evolution of the observable universe.

2. Mathematical Formulation and Observable Parameters

We start with the reparametrization-invariant action [5]

$$S = \int L d\tau = m \int d\tau \left[\frac{1}{2e} g_{\mu\nu} \dot{x}^\mu \dot{x}^\nu - \frac{e}{2} - \lambda g_{\mu\nu} \xi^\mu \dot{x}^\nu + \frac{e\lambda^2}{2} g_{\mu\nu} \xi^\mu \xi^\nu + \frac{e\beta\lambda^2}{2} \right]. \tag{1}$$

where τ is the worldline parameter, $\lambda(\tau)$ is an auxiliary scalar variable, e is the worldline einbein, β is a numerical constant. In a (4+1)-dimensional Schwarzschild-anti de Sitter (Sch-AdS) spacetime $g_{\mu\nu} dx^\mu dx^\nu = -\left(k - \frac{2M}{r^2} + \Lambda_5 r^2\right) dt^2 + \frac{dr^2}{k - \frac{2M}{r^2} + \Lambda_5 r^2} + r^2 d\Omega_3^2$, this action (1) becomes

$$S = \int L d\tau = \frac{m}{2} \int d\tau \, e \left[\frac{1}{e^2} \left(-f(r)\dot{t}^2 + \frac{\dot{r}^2}{f(r)} + r^2 \dot{\varphi}^2 \right) - \frac{2\lambda}{e} r^2 \dot{\varphi} + \lambda^2 (r^2 + \beta) - 1 \right], \tag{2}$$

where $f(r) = k - \frac{2M}{r^2} + \Lambda_5 r^2$, k is the curvature scalar and Λ_5 is the constant curvature of the space-time. Here, the energy and angular momentum of a test particle of mass m are conserved, which implies $p_t = -mF(r)\dot{t} = m\varepsilon, p_\varphi = mr^2(\dot{\varphi} - \lambda) = m\beta\lambda = ml$, where the overdot represents a proper-time derivative. The radial geodesic equation for (2) turns out to be

$$\dot{r}^2 + V_{eff}(r) = \varepsilon^2 \quad , \quad V_{eff} = F(r)\left(1 + l^2/r^2\right) \quad , \quad F(r) = \left(k - \frac{2M}{r^2} + \Lambda_5 r^2\right)\left(\frac{\mu^2 r^2 + l^2}{r^2 + l^2}\right), \tag{3}$$

which induces the $4+1$-dimensional effective metric of the form

$$ds_{4+1}^2 = -F(r)dt^2 + \frac{dr^2}{F(r)} + r^2 d\Omega_3^2. \tag{4}$$

Here μ ($\mu^2 = 1 + \frac{l^2}{\beta}$) is a dimensionless parameter. For $\mu^2 = 1$ (i.e. $\beta \longrightarrow \infty$), this reduces to standard Schwarzschild-AdS metric.

With this effective metric (4) in hand we now turn to its cosmological implications. We embed a $(3+1)$-dimensional FLRW space-time $(T, \sigma, \theta, \varphi)$:

$$ds_{3+1}^2 = -dT^2 + a(T)^2 \left[\frac{d\sigma^2}{1 - k\sigma^2} + \sigma^2 (d\theta^2 + \sin^2\theta \, d\varphi^2) \right], \tag{5}$$

into the $(4+1)$-dimensional effective metric (4) [5]. Employing the Gauss-Codazzi junction conditions and Z_2 symmetry [5, 4] and identifying $r(T)$ with $a(T)$, the Friedmann equations for spatially flat universe (k=0) takes the form:

$$\left(\frac{\dot{a}}{a}\right)^2 = \frac{8\pi G_4}{3}\rho + \frac{2M}{a^4} + \left[\alpha - \Lambda_5 + \frac{l^2(2M/a^2 - \Lambda_5 a^2)}{\beta(a^2 + l^2)}\right] \tag{6}$$

$$\frac{\ddot{a}}{a} = -\frac{4\pi G_4}{3}(\rho + 3p) - \frac{2M}{a^4} + \left[\alpha - \Lambda_5 - \frac{l^2(\Lambda_5 a^4 + 2M + 2\Lambda_5 l^2 a^2)}{\beta(a^2 + l^2)^2}\right] \tag{7}$$

where $\dot{a} = da/dT$, $8\pi G_4/3 = 2\rho_0 \left(8\pi G_5/3\right)^2$ and $\alpha = (8\pi G_5 \rho_0/3)^2$. The terms containing M ($\propto a^{-4}$) contributes to radiation energy density of the universe which is negligibly small for present day universe. Thus we express the Hubble parameter $H = \dot{a}/a$ in terms of redshift z ($1 + z = a_0/a = 1/a$) and neglect the radiation contributions (the terms of the order of $(1+z)^4$ or higher) to express the Friedmann equation (6) in a convenient form:

$$H^2 = H_0^2 \left[\Omega_X \left(1 + b(1+z)^2\right) + \Omega_M (1+z)^3\right] \tag{8}$$

where $\Omega_M = 8\pi G_4\rho/3H_0^2$ is the density parameter for the matter sector and $\Omega_X = (\alpha - \Lambda_5\mu^2)/H_0^2$ is the density parameter for the dark energy as induced by our generalized particle model. As we will see explicitly, the dimensionless parameter $b = \Lambda_5\beta(\mu^2-1)^2/(\alpha - \Lambda_5\mu^2)$, is responsible for the phantom behavior of the DE EOS. The most crucial part is to develop and estimate the observable parameters to show that we have a late accelerating universe where Ω_X accounts for the dark energy density. We also show that, consistent with observations, our candidate DE EOS (w_X) indeed crosses the phantom divider. As observations indicate a universe close to the ΛCDM model, this forces b to be small and β large (μ contains β in the denominator). These values can be restricted by χ^2 fitting.

Luminosity-redshift relation $d_L(z)$ determines dark energy density Ω_X from observations, and for our model it is given by

$$d_L(z) = (1+z)\int_0^z \frac{dz'}{H(z')} = \frac{(1+z)}{H_0}\int_0^z \frac{dz'}{\left[\Omega_X\left(1+b(1+z')^2\right)+\Omega_M(1+z')^3\right]^{1/2}}. \qquad (9)$$

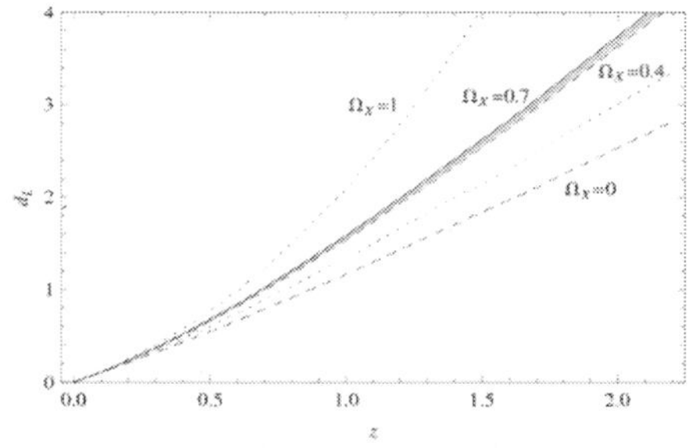

Figure 1. Variation of the luminosity distance with redshift for different Ω_X with $\Omega_M + \Omega_X = 1$

The plots in Figure 1 show that a universe with $\Omega_X \sim 0.7$, $\Omega_M \sim 0.3$ is favored, confirming that Ω_X accounts for the dark energy density. The allowed region (green shade) gives the bound for b as $-0.07 \leq b < 0$. Throughout the rest of the paper, we take a representative small negative value for $b = -0.05$. Once the luminosity distance is estimated, the apparent magnitude of the Supernovae can be calculated from the Hubble constant-free distance modulus $\mu(z) = 5\log_{10}[d_L(z)/\text{Mpc}] + 25$.

The deceleration parameter $q(z)$ explicitly shows the late accelerating behavior and for our model (8) we have

$$q(z) = \frac{-\ddot{a}/a}{\dot{a}^2/a^2} = \frac{H'(z)}{H(z)}(1+z) - 1 = \frac{\Omega_M(1+z)^3 - 2\Omega_X}{2(\Omega_X(1+b(1+z)^2)+\Omega_M(1+z)^3)}. \qquad (10)$$

The plot in Figure 2 confirms that our model indeed results in an early decelerating and late accelerating universe. Moreover, onset of the recent accelerating phase, when the universe was $\sim 60\%$ of its present size ($z = 0.6$), is also confirmed by our model.

The effective DE EOS in our model is given by

$$w_X(z) = \frac{2q(z)-1}{3[1-\Omega_M(z)]} = -1 + \frac{2b(1+z)^2}{3}. \qquad (11)$$

since b is negative and nonzero, it can be clearly seen from (11) that the effective EOS of dark energy candidate satisfies $w_X < -1$, exhibiting phantom behavior. Latest WMAP7 data

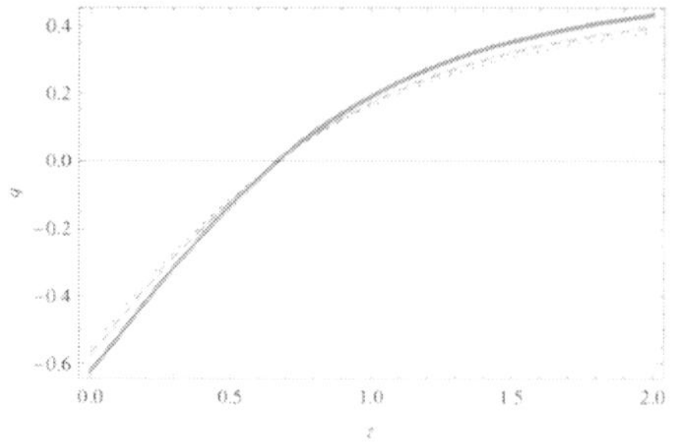

Figure 2. Variation of the deceleration parameter q with redshift

constrains the lower bound of the dark energy EOS to be $-1.12 < w_{DE}$ [2]. This sets the lower bound of b at $(-.16 \leq b)$, which is way below its lower bound $(-.07 \leq b)$ as predicted before from the luminosity-redshift relation. This model with $(-.07 \leq b < 0)$ will thus fit well with more precise bound for the EOS available in future.

The statefinder parameters $\{r, s\}$ [6] are defined as

$$r = \frac{\dddot{a}/a}{(\dot{a}/a)^3} = 1 + \left[\frac{H''}{H} + \left(\frac{H'}{H}\right)^2\right](1+z)^2 - 2\frac{H'}{H}(1+z) \; ; \; s = \frac{2}{3}\frac{r-1}{2q-1}$$

and can distinguish dynamical models from ΛCDM. In our model, $\{r, s\}$ pair is given by

$$r = 1 - \frac{b\Omega_X(1+z)^2}{\Omega_X(1+b(1+z)^2) + \Omega_M(1+z)^3} \; ; \; s = \frac{2}{3}\frac{b(1+z)^2}{[3+b(1+z)^2]}.$$

Hence, so far as b is non-trivial (as in the present case), the $\{r, s\}$ pair can distinguish our alternative gravity model from ΛCDM (for which $r = 1, s = 0$).

3. Summary and Future Prospects

To summarize, in our modified gravity scenario, induced by a higher dimensional non-minimal particle dynamics framework, one can have a phantom dark energy model *without the phantom*. Our model is qualitatively distinct, but not quantitatively far off, from ΛCDM model. Hence all the positive features of ΛCDM along with a phantom behavior (without the problems related to the negative kinetic term) can be accommodated in our model.

Several aspects of the proposed framework can be investigated further. The parameters used in the model can be constrained observationally by using maximum likelihood method involving the minimization of the function $\chi^2 = \sum_{i=1}^{N}\left[d_L(z)_{\rm obs} - d_L(z)_{\rm th}\right]^2/\sigma_i^2$, where $d_L(z)_{\rm th}$ contains the parameters used in a specific theory, N is the number of Supernovae (taken as 157 for the most reliable Gold dataset) and σ_i are the 1σ error from observational method used. Observationally, this is the most accurate probe of $\Omega_{\rm DE}$ and will further constrain the parameters used in our model. Further, the variable EOS may be reflected in Integrated Sachs-Wolfe (ISW) effect, which will serve as another test for the model. Studying features related to perturbations in this cosmological framework is another open issue.

[1] S. Perlmutter et al., *Astrophys. J.* **517** (1999) 565; A. Riess et al., *Astrophys. J.* **116** (1998) 1009
[2] http://lambda.gsfc.nasa.gov/product/map/dr4/parameters.cfm
[3] L. Perivolaropoulos, *JCAP* **0510** (2005) 001
[4] R. Maartens and K. Koyama, *Living Rev. Relativity* **13** (2010) 5; C. Barcelo and M. Visser, *Phys. Lett.* B **482** (2000) 183
[5] S. Das, S. Ghosh, J. W. van Holten and S. Pal, *JHEP* **0904** (2009) 115; *Int. J. Mod. Phys.* D **20** (2011) 1235
[6] U. Alam, V. Sahni, T. D. Saini and A. A. Starobinsky, *Mon. Not. Roy. Astron. Soc.* **344** (2003) 1057

Journal of Physics: Conference Series **405** (2012) 012013 doi:10.1088/1742-6596/405/1/012013

Exploring the Dark Ages and Epoch of Reionization with the HI 21 cm signal

Abhirup Datta

NASA Postdoctoral Fellow,Center for Astrophysics and Space Astronomy, University of Colorado, UCB 389, Boulder, CO-80309, USA

E-mail: abhirup.datta@colorado.edu

Abstract. Observations of the HI 21 cm transition line promises to be an important probe into the cosmic Dark Ages and Epoch of Reionization. Detection of this redshifted 21 cm signal is one of the key science goal for several upcoming and future low frequency radio telescopes like Murchison Widefield Array (MWA), Precision Array to Probe Epoch of Reionization (PAPER), LOw Frequency ARray (LOFAR) and Square Kilometer Array (SKA). One of the challenges for the detection of this signal is the accuracy of the foreground source removal. Here, we investigate the level of accuracy for the calibration and bright source removal algorithms from the reionization data-sets. We also put forward constraints for the design of the cosmic reionization data reduction scheme for the upcoming low frequency arrays such as, MWA, PAPER, etc. We show that the efficient foreground source removal strategies can only tolerate a frequency independent antenna based mean residual calibration error of 0.2% in amplitude or 0.2 degrees in phase. In order to extract the spherically averaged HI signal power spectrum, bright foregrounds need to be removed with calibration accuracies of 0.05% in amplitude. We also demonstrate the advantages of probing the cosmic reionization with the two dimensional HI power spectrum.

1. Introduction

The cosmological Epoch of Reionization (EoR) marks the transition from a fully neutral to a highly ionized intergalactic medium (IGM) due to the ultra-violet and X-ray radiation of early stars, galaxies, and black holes. Detection of the Gunn-Peterson effect, i.e., Lyα absorption by the neutral IGM, toward the most distant quasars ($z \sim 6$), and the large scale polarization of the CMB, corresponding to Thompson scattering during reionization, have set the first constraints on the reionization process. These results suggest significant variance in both space and time, starting perhaps as far back as $z \sim 11$ (WMAP 7 year data) and extending to $z \sim 6$ [3]. In case of the Gunn-Peterson effect, the IGM becomes optically thick to Lyα absorption for a neutral fraction as small as $\sim 10^{-3}$ at $z \sim 6$. Alternately, it has been widely recognized that mapping the red-shifted HI 21 cm line has great potential for direct studies of the neutral IGM during reionization [4,5].

There are number of upcoming low-frequency arrays with key science goals to detect the HI 21 cm signal from the EoR. This includes the Murchison Widefield Array [MWA],Precision Array to Probe Epoch of Reionization [PAPER], Low Frequency Array [LOFAR] and Giant Meterwave Radio Telescope [GMRT]. One of the major challenges for all of these upcoming arrays will be the removal of the continuum foreground sources in order to detect the faint HI signal from the EoR.

A variety of continuum foregrounds complicate redshifted 21 cm measurements of the EoR. Diffuse Galactic synchrotron emission dominates the low-frequency radio sky and is approximately four orders of magnitude brighter than the $\sim$ 10 mK HI 21 cm signal at the frequency relevant to reionization ($\nu \approx 150$ MHz). In addition, Galactic and extragalactic free-free emission contribute additional flux to the diffuse foreground. Radio point sources from AGN, radio galaxies, and local Galactic sources are numerous and particularly challenging. The brightest of these sources have fluxes well above 1 Jy and are seven or eight orders of magnitude above the EoR signal in low-frequency radio maps. The distribution of point sources also extends to very faint levels such that the brightness temperature due to confused sources in upcoming arrays will be $\sim$ 10 K, or three orders of magnitude brighter the than the 21 cm background.

In this paper we discuss how the radio interferometric imaging techniques are going to affect the foreground source modeling and subsequent removal from the data-set in order to search for the EoR signal. Extensive research is ongoing on foreground source modeling at these low frequencies and their subsequent removal from the EoR datasets. Since attempting to observe a signal below the confusion limit of foreground sources is a novel aspect of 21 cm experiments, most of these works primarily focus on the removal of faint and confused sources that fall below a specified cutoff flux limit of 1 Jy. Most of the other works implicitly assume that the bright foreground sources above 1 Jy have been removed perfectly from the raw EoR data-sets. In reality, imperfect instrument calibration or any errors in the subtracted foreground model will introduce artifacts and leave residual contamination in the data after bright source removal. These residuals may interfere with either the subsequent faint source subtraction or the ultimate detection and characterization of the redshifted HI 21 cm signal.

In [1] we dealt with removal of the bright point sources above 1 Jy and the limitations that will be caused due to imperfect removal of such sources in the image plane. In [2] we have extended these initial analysis in order to estimate the residual contamination in the power spectral domain of improper bright source subtraction.

2. Method of Simulation

For our simulations, we have used a simple sky model that only includes bright radio point sources. No diffuse emission from the Galaxy is included as a part of the sky model; and the 21 cm signal and thermal noise are also omitted. Our sky model is derived from the $\log N$–$\log S$ distribution of source counts from the 6C survey at 151 MHz.

We have assumed a fiducial MWA configuration for these simulations. The details of the array parameters are included in Table 1. For a field-of-view of 15°, total number of point sources above 1 Jy is $\sim$ 170. In our sky model all the foreground sources are flat spectrum, i.e. with zero spectral index. Hence, the only frequency dependent contribution is from the point spread function or the array beam.

The observed visibilities are simulated for a 6 hour observation ($\pm$3 hours in Hour-angle) using the sky model and the array specifications. Next, we generate the foreground model corrupted with calibration errors that will be subtracted from the observed visibilities. Here, we assume that the residual errors in a given antenna are perfectly correlated for the duration of one 6-hour observing night, but perfectly uncorrelated between successive observing nights. We further assume that the residual errors *between* antennas are perfectly uncorrelated at all times. The residual visibilities are calculated by subtracting the model visibilities from the simulated visibilities. Resultant images from the above steps followed in the simulations are shown in panels (a) of Figures 5, 7, and 9 in [1]. Now, intermediate image cubes are obtained from these intermediate residual visibilities and are subjected to the second stage of redshifted 21 cm foreground subtraction that aims to remove faint and confused sources by fitting and subtracting a low-order polynomial along the frequency axis for each line of sight in the data cube. Thus we obtain the final residual image cubes (as shown in panels (b) of Figures 5, 7,

Table 1. Array Specifications

Parameters	Values
No. of Tiles	512
Central Frequency	158 MHz ($z \sim 8$)
Field of View	$\sim 15^o$ at 158 MHz. ($\propto \lambda$)
Synthesized beam	$\sim$ 4.5' at 158 MHz. ($\propto \lambda$)
Effective Area per Tile	$\sim 17\ m^2$
Maximum Baseline	$\sim$ 1.5 km
Total Bandwidth	32 MHz
T_{sys}	$\sim$ 250 K
Channel Width	$\sim$ 32 kHz
Thermal Noise	$\sim$ 7.55 mK (5000 hours & 2.5 MHz)

and 9 in [1]).

The final step in our simulations is to calculate power spectra from the intermediate and final residual image cubes. These residual foreground power spectra are compared with the theoretically predicted 21 cm power spectrum and expected thermal noise power spectrum for the MWA. We calculate three forms of the residual power spectra from our final data cubes: the derived angular power spectrum for a narrow frequency channel (shown as Figure 6 in [2]), the spherically averaged three-dimensional power spectrum from the entire data cube(shown as Figure 8 in [2]), and the two-dimensional power spectrum found by averaging over transverse modes in the full three-dimensional power spectrum(shown as Figure 11 in [2]).

In order to simulate the observed visibilities, we have used the simulator tool in the CASA software. We have also used CASA to perform the imaging, visibility subtraction and polynomial subtraction. The rest of the operations are performed using separately written Python scripts.

3. Results and Conclusions

In [1], we explored the calibration accuracy needed to allow direct imaging of cosmic Stromgren spheres with very deep integrations by the MWA. Our simulations demonstrated that a calibration accuracy of 0.2% systematic error in gain amplitude per night of observing (or 0.2 degree in phase) is needed for the residual contamination to be below the thermal noise in a part of the image map far from any bright sources for a long integration by the MWA.In other words, the above accuracies will allow detection of HI 21 cm features from EoR by direct imaging using the MWA.

In [2], we focus our attention on the residual contamination that can be tolerated in measurements of HI 21 cm power spectra, rather than direct imaging of the HI 21 cm background. The information in the HI 21 cm signal field is expected to be effectively compressed in a power spectrum by averaging over many Fourier modes. Hence, the signal-to-noise ratio (SNR) in a properly chosen power spectrum measurement can be significantly higher than the per-pixel SNR in a direct imaging observation. In fact, detection of the HI 21 cm power spectrum can be achieved with the sensitivity of first generation EoR telescopes like MWA, PAPER and LOFAR. But for direct imaging of the HI 21 cm background we might need the sensitivity of next generation telescopes like the SKA.

The spherically-averaged three-dimensional 21 cm power spectrum is the primary reionization observable targeted by the MWA. Our results show that the residual calibration accuracy of < 0.05 % would allow the detection of HI 21 cm signal in spherically-averaged three-dimensional power spectrum (Figure 8 in [2]). In comparison to the angular power spectrum, we infer that the spherically-averaged power spectra has a better tolerance for the residual calibration errors. This also reflects the fact that the angular power spectrum has been produced using a single channel map of 125 kHz, whereas the spherically-averaged spectrum is produced with the total bandwidth of 32 MHz.

However, the spherically-averaged power spectrum mixes the contribution from the line-of-sight and the transverse modes of the three-dimensional power spectrum. It should be noted that both the foregrounds and the predicted redshifted 21 cm signal that includes redshifted-space distortions have *aspherical* structure in the Fourier domain [6]. Hence, it is critical to decouple the contributions of these two modes by constructing the two-dimensional power spectrum. In Figure 1, we have shown the results of the obtained two-dimensional power spectra from the final residual visibilities (for 0.1% calibration error). Figure 2 shows the expected HI 21 cm power spectrum. These results show clear advantages of the two-dimensional power spectrum over the spherically-averaged power spectrum. The results for residual calibration errors indicate that at transverse $k < 0.05$ Mpc^{-1}, we are able to probe most of the line-of-sight k scales where the HI signal is dominant over the residual foreground power.

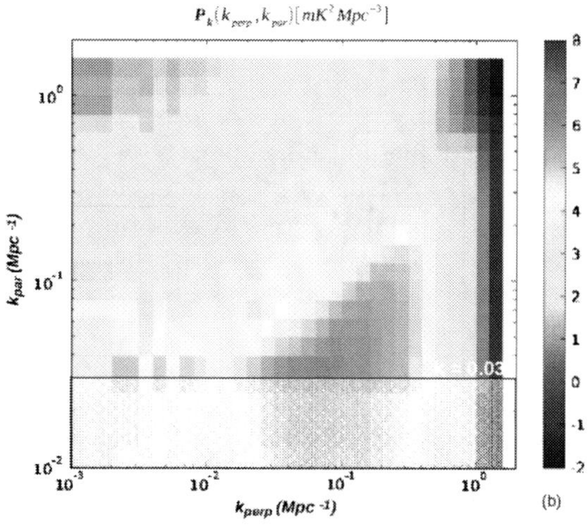

Figure 1. The two-dimensional power spectrum from the final residual image cube is shown for a fiducial calibration error level of 0.1% (as in Figure 11-b of [2]).

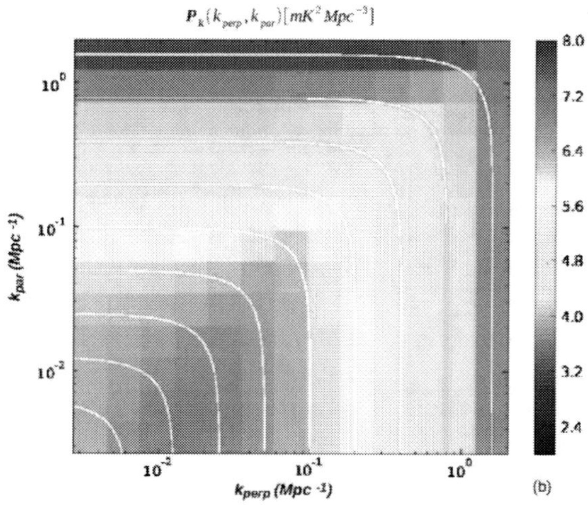

Figure 2. Predicted theoretical two-dimensional power spectrum of the HI 21 cm signal (as in Figure 9-b of [2]).

We expect to build on our present analysis in future work by exploring other arrays like LOFAR or SKA, addressing foreground removal issues of extended bright sources and including calibration issues such as wide-field gain calibration of the primary beam and the ionosphere.

References

[1] Datta A, Bhatnagar S and Carilli C L 2009,*ApJ*,**703**, 1851
[2] Datta A, Bowman J D and Carilli C L 2010,*ApJ*, **724**, 526
[3] Fan X, Strauss M A, Becker R H, White R L, Gunn J E, Knapp G R, Richards G T, Schneider D P, Brinkmann J and Fukugita M 2006, *AJ*, **132**, 117
[4] Furlanetto S R, Oh S P and Briggs F H 2006, *Phys. Rep.*, **433**, 181
[5] Morales M F and Wyithe J S B 2010, *ARA&A*, **48**, 127
[6] Morales M F, Bowman J D and Hewitt J N 2006,*ApJ*, **648**, 767

Journal of Physics: Conference Series **405** (2012) 012014 doi:10.1088/1742-6596/405/1/012014

Noncommutative inspired Schwarzschild black hole, Voros product and Komar energy[1]

Sunandan Gangopadhyay

Department of Physics, West Bengal State University, Barasat, Kolkata 700126, India

E-mail: sunandan.gangopadhyay@gmail.com

Abstract. The importance of the Voros product in defining a noncommutative Schwarzschild black hole is shown. The entropy is then computed and the area law is shown to hold upto order $\frac{1}{\sqrt{\theta}}e^{-M^2/\theta}$. The leading correction to the entropy (computed in the tunneling formalism) is shown to be logarithmic. The Komar energy E for these black holes is then obtained and a deformation from the conventional identity $E = 2ST_H$ is found at the order $\sqrt{\theta}e^{-M^2/\theta}$. This deformation leads to a nonvanishing Komar energy at the extremal point $T_H = 0$ of these black holes. Finally, the Smarr formula is worked out. Similar features also exist for a deSitter–Schwarzschild geometry.

In this presentation, we discuss some of the issues of noncommutative inspired Schwarzschild black hole [3, 4] which has gained considerable interest recently. The primary point of interest is that there is no clear cut connection of this type of noncommutativity with standard notions of a noncommutative (NC) spacetime where point-wise multiplications are replaced by appropriate star multiplications. We shall point out that the Voros star product [5] plays an important role in obtaining the mass density of a static, spherically symmetric, smeared, particle-like gravitational source. Our second objective is to derive quantum corrections to the semiclassical Hawking temperature and entropy by the tunneling mechanism by going beyond the standard semiclassical approximation [6] for this black hole spacetime. Finally, we would like to review the status of the relation between the Komar energy [7], entropy and Hawking temperature ($E = 2ST_H$) in the context of NC inspired Schwarzschild black holes.

The first issue can be addressed by taking recourse to the formulation and interpretational aspects of NC quantum mechanics [8, 9]. We observe that the inner product of the coherent states $|z, \bar{z})$ (used in the construction of the wave-function of a free point particle) can be computed by using a deformed completeness relation (involving the Voros product) among the coherent states

$$\int \frac{\theta dz d\bar{z}}{2\pi}\ |z, \bar{z}) \star (z, \bar{z}| = 1_Q \tag{1}$$

where the Voros star product between two functions $f(z, \bar{z})$ and $g(z, \bar{z})$ is defined as

$$f(z, \bar{z}) \star g(z, \bar{z}) = f(z, \bar{z}) e^{\overleftarrow{\partial_{\bar{z}}}\overrightarrow{\partial_z}} g(z, \bar{z})\ . \tag{2}$$

[1] This presentation is based on the work in references [1, 2].

Published under licence by IOP Publishing Ltd

The wave-function of the free point particle on the NC plane [10, 8] is given by

$$\psi_{\vec{p}} = (p|z,\bar{z}) = \frac{1}{\sqrt{2\pi\hbar^2}} e^{-\frac{\theta}{4\hbar^2}\bar{p}p} e^{i\sqrt{\frac{\theta}{2\hbar^2}}(p\bar{z}+\bar{p}z)} \quad ; \quad p = p_x + ip_y\ ,\ z = \frac{1}{\sqrt{2\theta}}(x+iy) \tag{3}$$

where the momentum eigenstates are normalised such that $(p'|p) = \delta(p'-p)$ and satisfy the completeness relation

$$\int d^2p\ |p)(p| = 1_Q\ . \tag{4}$$

It turns out that a consistent probabilistic interpretation of this wave-function can be given only when the Voros product is incorporated. With these observations and interpretations in place, we now write down the overlap of two coherent states $|\xi,\bar{\xi})$ and $|w,\bar{w})$ using the completeness relation for the position eigenstates in eq.(1)

$$(w,\bar{w}|\xi,\bar{\xi}) = \int \frac{\theta dz d\bar{z}}{2\pi}\ (w,\bar{w}|z,\bar{z}) \star (z,\bar{z}|\xi,\bar{\xi})\ . \tag{5}$$

A simple inspection shows that $(w,\bar{w}|z,\bar{z}) = \frac{1}{\theta}e^{-|\omega-z|^2}$ satisfies the above equation. A straightforward dimensional lift of this solution from two to three space dimensions immediately motivates one to write down the mass density of a static, spherically symmetric, smeared, particle-like gravitational source in three space dimensions as

$$\rho_\theta(r) = \frac{M}{(4\pi\theta)^{3/2}} \exp\left(-\frac{r^2}{4\theta}\right)\ . \tag{6}$$

The above discussion clearly brings out the important role played by the Voros product in defining the mass density of the NC Schwarzschild black hole. Solving Einstein's equations with the above mass density incorporated in the energy-momentum tensor leads to the following NC inspired Schwarzschild metric [3],[4]

$$ds^2 = -\left(1 - \frac{4M}{r\sqrt{\pi}}\gamma(\frac{3}{2},\frac{r^2}{4\theta})\right)dt^2 + \left(1 - \frac{4M}{r\sqrt{\pi}}\gamma(\frac{3}{2},\frac{r^2}{4\theta})\right)^{-1} dr^2 + r^2(d\tilde{\theta}^2 + \sin^2\tilde{\theta}d\phi^2)\ . \tag{7}$$

The event horizon of the black hole can be found by setting $g_{tt}(r_h) = 0$ in eq.(7), which yields

$$r_h = \frac{4M}{\sqrt{\pi}}\gamma(\frac{3}{2},\frac{r_h^2}{4\theta}). \tag{8}$$

The large radius regime ($\frac{r_h^2}{4\theta} >> 1$) is taken where one can expand the incomplete gamma function to solve r_h by iteration. Keeping upto next to leading order $\sqrt{\theta}e^{-M^2/\theta}$, we find

$$r_h \simeq 2M\left[1 - \frac{2M}{\sqrt{\pi\theta}}\left(1+\frac{\theta}{2M^2}\right)e^{-M^2/\theta}\right]\ . \tag{9}$$

Now for a general stationary, static and spherically symmetric space time, the Hawking temperature (T_H) is related to the surface gravity (κ) by the following relation [11]

$$T_H = \frac{\kappa}{2\pi} \tag{10}$$

Journal of Physics: Conference Series **405** (2012) 012014 doi:10.1088/1742-6596/405/1/012014

where the surface gravity of the black hole is given by

$$\kappa = \frac{1}{2}\left[\frac{dg_{tt}}{dr}\right]_{r=r_h} . \tag{11}$$

Hence the Hawking temperature for the NC inspired Schwarzschild black hole upto order $\sqrt{\theta}e^{-M^2/\theta}$ is given by

$$T_H = \frac{1}{8\pi M}\left[1 - \frac{4M^3}{\theta\sqrt{\pi\theta}}\left(1 - \frac{\theta}{2M^2} - \frac{\theta^2}{4M^4}\right)e^{-M^2/\theta}\right] . \tag{12}$$

The first law of black hole thermodynamics can now be used to work out the Bekenstein-Hawking entropy. The law reads

$$dS_{BH} = \frac{dM}{T_H} . \tag{13}$$

Hence the Bekenstein-Hawking entropy in the next to leading order in θ is found to be

$$S_{BH} = \int \frac{dM}{T_H} = 4\pi M^2 - 16\sqrt{\frac{\pi}{\theta}}M^3\left(1 + \frac{\theta}{M^2}\right)e^{-M^2/\theta} . \tag{14}$$

To express the entropy in terms of the NC horizon area (A_θ), we use eq.(9) to get

$$A_\theta = 4\pi r_h^2 = 16\pi M^2 - 64\sqrt{\frac{\pi}{\theta}}M^3\left(1 + \frac{\theta}{2M^2}\right)e^{-M^2/\theta} + \mathcal{O}(\theta^{3/2}e^{-M^2/\theta}). \tag{15}$$

Comparing equations (14) and (15), we find that at the leading order in θ (i.e. upto order $\frac{1}{\sqrt{\theta}}e^{-M^2/\theta}$), the NC black hole entropy satisfies the area law (in the regime $\frac{r_h^2}{4\theta} >> 1$)

$$S_{BH} = \frac{A_\theta}{4} . \tag{16}$$

We now look for corrections to the semiclassical area law upto leading order in θ.
To do so, we first compute the corrected Hawking temperature $\tilde{T}_H$. For that we use the tunneling method by going beyond the semiclassical approximation [6]. Considering the massless scalar particle tunneling under the background metric (7), the corrected Hawking temperature is given by

$$\tilde{T}_H = T_H\left[1 + \sum_i \frac{\tilde{\beta}_i \hbar^i}{(Mr_h)^i}\right]^{-1} . \tag{17}$$

Applying the first law of black hole thermodynamics once again with this corrected Hawking temperature, gives the following expression for the corrected entropy/area law :

$$\begin{aligned} S &= \frac{A_\theta}{4\hbar} + 2\pi\tilde{\beta}_1 \ln A_\theta - \frac{64\pi^2\tilde{\beta}_2\hbar^2}{A_\theta} + \mathcal{O}(\sqrt{\theta}e^{-\frac{M^2}{\theta}}) \\ &= S_{BH} + 2\pi\tilde{\beta}_1 \ln S_{BH} - \frac{16\pi^2\tilde{\beta}_2\hbar}{S_{BH}} + \mathcal{O}(\sqrt{\theta}e^{-\frac{M^2}{\theta}}) . \end{aligned} \tag{18}$$

Finally, we proceed to investigate the status of the relation between the Komar energy E, entropy S and Hawking temperature T_H

$$E = 2ST_H \tag{19}$$

Journal of Physics: Conference Series **405** (2012) 012014 doi:10.1088/1742-6596/405/1/012014

in the case of these NC inspired black holes. The expression for the Komar energy E for the NC inspired Schwarzschild metric (7) is given by [1]

$$E = \frac{2M}{\sqrt{\pi}}\gamma\left(\frac{3}{2}, \frac{r^2}{4\theta}\right) - \frac{Mr^3}{2\theta\sqrt{\pi\theta}}e^{-r^2/(4\theta)} . \tag{20}$$

We therefore identify M as the mass of the black hole since $E = M$ in the limit $r \to \infty$. This identification plays an important role as we shall see below.
The above expression computed near the event horizon of the black hole[2] upto order $\sqrt{\theta}e^{-M^2/\theta}$ gives

$$E = M\left[1 - \frac{2M}{\sqrt{\pi\theta}}\left(\frac{2M^2}{\theta} + 1\right)e^{-M^2/\theta} - \frac{1}{M}\sqrt{\frac{\theta}{\pi}}e^{-M^2/\theta}\right] . \tag{21}$$

Finally, using eqs.(12), (14) and (21), we obtain

$$\begin{aligned} E &= 2ST_H + 2\sqrt{\frac{\theta}{\pi}}e^{-M^2/\theta} + \mathcal{O}(\theta^{3/2}e^{-M^2/\theta}) \\ &= 2ST_H + 2\sqrt{\frac{\theta}{\pi}}e^{-S/(4\pi\theta)} + \mathcal{O}(\theta^{3/2}e^{-S/(4\pi\theta)}) \end{aligned} \tag{22}$$

where in the second line we have used eq.(14) to replace M^2 by $S/(4\pi)$ in the exponent. Interestingly, we find that the relation $E = 2ST_H$ gets deformed upto order $\sqrt{\theta}e^{-M^2/\theta}$ which is consistent with the fact that the area law also gets modified at this order. The deformation also yields a nonvanishing Komar energy at the extremal point $T_H = 0$ of these black holes [2]. Also, we have once again managed to write down the deformed relation in terms of the Komar energy E, entropy S and the Hawking temperature T_H. Similar features are also present for a de-Sitter Schwarzschild geometry [12]. Eq.(22) can also be written with M being expressed in terms of the black hole parameters S and T_H using eq.(21)

$$M = 2ST_H + \frac{1}{2\pi\sqrt{\pi\theta}}\left(S + \frac{S^2}{2\pi\theta} + 6\pi\theta\right)e^{-S/(4\pi\theta)} + \mathcal{O}(\theta^{3/2}e^{-S/(4\pi\theta)}). \tag{23}$$

We name eq.(23) as the *Smarr formula* [13] *for NC inspired Schwarzschild black hole* since M has been identified earlier to be the mass of the black hole.

[1] Banerjee R, Gangopadhyay S and Modak S K 2010 Phys. Lett. B **686** 181 [arXiv:0911.2123 [hep-th]].
[2] Banerjee R and Gangopadhyay S 2011 Gen. Rel. Grav **43** 3201 [arXiv:1008.1683 [hep-th]].
[3] Nicolini P, Smailagic A and Spallucci E 2006 Phys. Lett. B **632** 547 [arXiv:gr-qc/0510112].
[4] Nicolini P 2009 Int.J.Mod.Phys.A **24** 1229 [arXiv:0807.1939 [hep-th]].
[5] Voros A 1989 Phys. Rev. A **40** 6814.
[6] Banerjee R and Majhi B R 2008 JHEP **06** 095 [arXiv:0805.2220].
[7] Komar A 1959 Phys. Rev. **113** 934.
[8] Scholtz F G, Gouba L, Hafver A and Rohwer C M 2009 J. Phys. A **42** 175303.
[9] Gangopadhyay S and Scholtz F G 2009 Phys. Rev. Lett. **102** 241602.
[10] Smailagic A and Spallucci E 2003 J. Phys. A **36** L467.
[11] Bardeen J M, Carter B and Hawking S W 1973 Comm.Math.Phys. **31** 161.
[12] Dymnikova I 1992 Gen. Rel. Grav. **24** 235.
[13] Smarr L 1973 Phys. Rev. Lett. **30** 71.

[2] The Komar energy computed near the event horizon of the black hole plays an important role in obtaining the coefficient of the logarithmic correction term (18) in the entropy [1].

International Conference on Modern Perspectives of Cosmology and Gravitation (COSGRAV12) IOP Publishing
Journal of Physics: Conference Series **405** (2012) 012015 doi:10.1088/1742-6596/405/1/012015

A discussion on Lorentz preserving scalar fields in Lorentz violating theory

O Ganguly[1], **D Gangopadhyay**[2], **P Majumdar**[3]
[1] S. N. Bose National Centre for Basic Sciences, Kolkata 700098, India
[2] Ramakrishna Mission Vivekananda University, Howrah, West Bengal 711202, India
[3] Saha Institute of Nuclear Physics, Kolkata 700064, India

E-mail: oindrila@bose.res.in, debashis@rkmvu.ac.in, parthasarathi.majumdar@saha.ac.in

Abstract. Lorentz symmetry can be preserved in effective higher derivative scalar field theories containing a constant vector that breaks Lorentz invariance of flat spacetime, through the choice of special field configurations. These fields do satisfy the equations of motion, yielding cubic dispersion relations analogous to those derived earlier. Moreover, the Lie algebra of the Lorentz group can be realised on these fields.

1. Introduction

Physicists have focussed a lot of their attention on theoretically and experimentally probing departures from Lorentz symmetry for the better part of the last couple of decades, [1, 2, 3, 4, 5, 6, 7, 8, 9]. This search was initiated by different theories of quantum gravity that hypothesise violation of Lorentz symmetry in flat spacetime at energies $E \gg M_{Pl}$, where M_{Pl} denotes the Planck mass. Deviations, suppressed by the Planck mass, from the standard special relativistic dispersion relation of free particles of mass m at high energies E ($M_{Pl} \gg E \gg m$) are considered to be signatures of quantum gravity induced Lorentz violation. Myers and Pospelov [6] constructed low energy effective actions of fields having spins 0, 1/2 and 1 that include new Planck suppressed dimension five operators, to account for these proposed corrections to the dispersion relations. In this report, we shall restrict ourselves to the case of a complex scalar field ϕ. The extended action of [6] is,

$$S_{MP_\phi} = \int d^4x \mathcal{L}_{MP_\phi} = \int d^4x \left[|\partial\phi|^2 - m^2|\phi|^2 \right] + \int d^4x \frac{i\kappa}{M_{Pl}} \phi^* \partial_n^3 \phi \ , \tag{1}$$

The first integral, let's call it S_S, is the usual action of a complex scalar field with mass m while the second integral, S_{V_S}, is the Lorentz violating contribution. Global Lorentz symmetry of flat spacetime is broken by the constant vector $\mathbf{n}$ which renders a preferred direction to the background. κ is a real, dimensionless parameter and $\mathbf{n} \cdot \partial \equiv \partial_n$. Choosing a Lorentz frame where $n^\mu = (1, \vec{0})$, corrections of $O(p^3)$ to the dispersion relation has been obtained in [6] in the limit $M_{Pl} \gg E \gg m$.

$$\omega^2 \simeq |\vec{p}|^2 + m^2 + \frac{\kappa}{M_{Pl}} |\vec{p}|^3 \ . \tag{2}$$

Although such deformed dispersion relations have attracted much attention, do they unequivocally imply Lorentz violation? We first explore the possibility that special field configurations exist for which the apparently Lorentz symmetry violating action of [6] may still be Lorentz *invariant* in close analogy to what happens in magnetic monopole theory [10]. In section 2 we consider the Nöther current corresponding to Lorentz transformation of the action (1). Demanding that this Nöther current be conserved leads us to the the result that the effective action remains Lorentz invariant if the fields are decomposed in a particular way. The initial absence of Lorentz symmetry in the action is transferred to the Lorentz non-invariant splitting of the fields. We further investigate in the same section, the dispersion relation that these special field configurations satisfy. This work is based on our more extensive study [11]

One must also have noticed that the additional term S_{V_S} in (1) contains third order derivatives of the field. Higher derivative Lagrangians are not new to physics [12]. Back in 1961, Ostrogradskii [13] had developed a canonical formalism for dealing with them. However, most of the extant literature in this field, [14, 15, 16] and references therein, study systems having finite number of degrees of freedom and higher time derivatives of the generalised coordinates. Section 3 will present a brief review of Ostrogradskii's technique generalised to field systems [17] and use it to study the canonical structure of $\mathcal{L}_{MP_\phi}$ with the Lorentz-preserving fields. This part of the problem is going to be dealt with in more detail in our forthcoming publication [18].

2. Lorentz invariant dynamics of Myers Pospelov scalar model

2.1. *Finding the Lorentz preserving fields*

Let an infinitesimal Lorentz transformation be applied to the action (1). Then $\delta_{\alpha\beta}S_S = 0$ while $\delta_{\alpha\beta}S_{V_S} = \int d^4x\phi^* n_{[\alpha}\partial_{\beta]}\partial_n^2\phi$. On the other hand, if the spacetime divergence of the corresponding Nöther current $\mathcal{J}$ is computed, we get $\partial_\mu \mathcal{J}^\mu_{\alpha\beta} = \phi^* n_{[\alpha}\partial_{\beta]}\partial_n^2\phi$.

Lorentz transformation will be a symmetry of the system if both $\delta_{\alpha\beta}S_{V_S} = 0$ and $\partial_\mu \mathcal{J}^\mu_{\alpha\beta} = 0$. This yields the condition $n_{[\alpha}\partial_{\beta]}\partial_n^2\Phi = 0$. A possible non-trivial solution is,

$$\phi(\mathbf{x}) = \phi_\parallel(\mathbf{x}_\parallel) + \phi_\perp(\mathbf{x}_\perp) \,, \tag{3}$$

where $\phi_\parallel$ and $\phi_\perp$ are arbitrary functions of their respective arguments $\mathbf{x}_\parallel$ and $\mathbf{x}_\perp$, defined by $\mathbf{x}_\parallel \equiv \frac{\mathbf{x}\cdot\mathbf{n}}{n^2}\mathbf{n}$ and $\mathbf{n}\cdot\mathbf{x}_\perp = 0$. So, $\mathbf{x} = \mathbf{x}_\parallel + \mathbf{x}_\perp$. Hence, the requirement of a Lorentz invariant action imposes a non-trivial restriction on the functional form of fields. Now, if we choose $\mathbf{n}$ to be timelike, we can align x^0 along $\mathbf{n}$. Then our condition (3) implies that when the full scalar field is a linear combination of a time-dependent, spatially homogeneous piece and a static spatially inhomogeneous piece, *the theory possesses Lorentz symmetry.*

2.2. *Evaluation of dispersion relation*

The scalar field $\phi(x)$ assumed to be given by (3) leads to the equation of motion $(\Box + m^2)\phi = \frac{i\kappa}{M_{Pl}}\partial_n^3\phi$ to be written as $(\nabla_\perp^2 + m^2)\phi_\perp = -(\nabla_\parallel^2 + m^2)\phi_\parallel + \frac{i\kappa}{M_{Pl}}\partial_n^3\phi_\parallel$, for $\mathbf{n}$ a unit vector (we have used the decomposition $\Box\phi = \nabla_\parallel^2\phi_\parallel + \nabla_\perp^2\phi_\perp$). By taking the simple ansätz $\phi_\perp \sim \exp(-\mathbf{k}_\perp\cdot\mathbf{x}_\perp)$, $\phi_\parallel \sim \exp(-i\,\mathbf{k}_\parallel\cdot\mathbf{x}_\parallel)$ and going over to the inertial frame characterised by $\mathbf{n} = (1, \vec{0})$, it can be shown that the dispersion relation of the complete scalar field $\phi(\mathbf{x})$ in the high energy regime $\mathcal{E} \simeq |\vec{k}| \gg m$ takes the simple form,

$$\mathcal{E}^2 \simeq |\vec{k}|^2 + \frac{\kappa}{M_{Pl}}\,|\vec{k}|^3, \tag{4}$$

provided the four momentum $\mathbf{k} = (\mathcal{E}, \vec{k})$. This is same as the dispersion relation (2) computed in [6]. Hence, observation of a deformed dispersion relation doesn't necessarily guarantee the existence of Lorentz violating dynamics.

3. Canonical structure of Lorentz preserving Lagrangian

For the rest of this report, we will study the canonical structure of Myers Pspelov theory in the presence of Lorentz preserving fields.

3.1. Canonical formalism in higher derivative field theories

Let us consider a system of scalar fields $\phi_a(\mathbf{x})$ described by the Lagrangian density $\mathcal{L}(\phi_a, \phi_{a,\rho_1}, \phi_{a,\rho_1\rho_2}, ..., \phi_{a,\rho_1...\rho_l})$. Here, $\phi_{a,\rho_1...\rho_j} \equiv \partial_{\rho_1}...\partial_{\rho_j}\phi_a$ and a labels the different fields. A canonical formalism of relativistic field theories requires us to 'foliate' spacetime into constant time, spacelike hypersurfaces. This in turn involves the separation of temporal and spatial derivatives of the fields[1]. It is most often possible to arrange terms in the Lagrangian such that "mixed" derivatives of the fields, as in $\partial_t^2\partial_k\phi$, do not survive. This is true not only when the Lagrangian is Lorentz invariant [17] but also when the Lorentz violating action (1) is written in terms of the Lorentz preserving fields [18]. In such situations, the canonical momenta are given by

$$\pi^{a(j)} \equiv \sum_{i=j+1}^{l} (-d_t)^{i-(j+1)} \frac{\partial \mathcal{L}}{\partial \phi_{a(i)}}, \quad j = 0, ..., l-1 \,. \tag{5}$$

We have adopted the notation $\phi_{a(j)} \equiv d_t^j \phi_a$, $j = 0, ..., l$ and $\pi^{a(j)}$ are the momenta conjugate to $\phi_{a(j)}$ upto $j = l - 1$. It is evident that we are now working in an enlarged phase space. The canonical variables satisfy the Poisson bracket $\{\phi_{a(i)}(t, \vec{x}), \pi^{b(j)}(t, \vec{x}')\} = \delta_a^b\, \delta_i^j\, \delta^{(3)}(\vec{x} - \vec{x}')$. We also wish to stress that higher derivative theories owe their peculiarities only to higher order time derivatives of the fields, spatial derivatives being rather benign.

3.2. Algebra of Lorentz preserving fields

Without loss of generality, we are going to work in a Lorentz frame defined by $\mathbf{n} = (1, \vec{0})$. This greatly simplifies the Lagrangian density:

$$\mathcal{L}_{MP_\phi} = \dot{\phi}_\parallel^* \dot{\phi}_\parallel - \vec{\nabla}\phi_\perp^* \cdot \vec{\nabla}\phi_\perp + \frac{i\kappa}{M_{Pl}}(\phi_\parallel^* + \phi_\perp^*)\dddot{\phi}_\parallel \,. \tag{6}$$

Here, we have neglected the masses of the fields as we are interested in behaviour of the system at energies much higher than the field masses. Eq.(6) has only higher order time derivatives of $\phi_\parallel(t)$ while $\vec{\nabla}\phi_\parallel = 0 = \partial_t\phi_\perp$. Thus all mixed derivatives in the sense described above will vanish. This permits us to safely use eq.(5) to determine the canonical momenta. We list them in table 1.

The Nöther charge $Q_{\alpha\beta} = \int d\sigma\ \sigma_{\rho_1} \mathcal{J}^{\rho_1}_{\alpha\beta}$, Σ being a three dimensional hypersurface. If we orient it orthogonal to the time axis then the Nöther charge may be written in terms of the canonical variables:

$$Q_{\alpha\beta} = \int d^3\vec{x} \left(\pi_\parallel^{(0)} \delta_{\alpha\beta}\phi_{\parallel(0)} + \pi_\parallel^{(1)} \delta_{\alpha\beta}\phi_{\parallel(1)} + \pi_\parallel^{(2)} \delta_{\alpha\beta}\phi_{\parallel(2)} + \pi_{\parallel^*}\delta_{\alpha\beta}\phi_\parallel^* - x_{[\alpha}\delta^0_{\beta]}\mathcal{L}_{MP_\phi} \right) . \tag{7}$$

From here we can deduce the action of the Nöther charge on the Lorentz preserving fields and the algebra satisfied by them [18]. The results appear below in eqs. (8), (9), (10)

[1] At this stage, it is imperative that we sort the indices. Latin letters from the middle of the alphabet set viz. i,j,k,l are being used as summation indices while those from the end like r,s,...,z will denote spatial components. Different fields will be labelled by the alphabets a,b,c,d. Greek letters are reserved for spacetime indices.

Generalised coordinate	Generalised momentum
$\phi_{\parallel(0)} \equiv \phi_{\parallel}$	$\pi_{\parallel(0)} = \dot{\phi}^*_{\parallel} + \frac{i\kappa}{M_{Pl}}\ddot{\phi}^*_{\parallel}$
$\phi_{\parallel(1)} \equiv \dot{\phi}_{\parallel}$	$\pi_{\parallel(1)} = -\frac{i\kappa}{M_{Pl}}\dot{\phi}^*_{\parallel}$
$\phi_{\parallel(2)} \equiv \ddot{\phi}_{\parallel}$	$\pi_{\parallel(2)} = \frac{i\kappa}{M_{Pl}}(\phi^*_{\parallel} + \phi^*_{\perp})$
$\phi^*_{\parallel(0)} \equiv \phi^*_{\parallel}$	$\pi_{\parallel^*(0)} = \dot{\phi}_{\parallel}$
$\phi_{\perp(0)} \equiv \phi_{\perp}$	$\pi_{\perp(0)} = 0$
$\phi^*_{\perp(0)} \equiv \phi^*_{\perp}$	$\pi_{\perp^*(0)} = 0$

Table 1. Canonically conjugate phase space variables

$$\{Q_{\alpha\beta}(t), \phi_{b(k)}(t, \vec{x})\} = -\delta_{\alpha\beta}\phi_{b(k)}(t, \vec{x}), \quad for\ \phi_b = \phi_{\parallel}, \phi^*_{\parallel} , \tag{8}$$

$$= 0, \quad for\ \phi_b = \phi_{\perp}, \phi^*_{\perp} , \tag{9}$$

$$\{Q_{\alpha\beta}(t), Q_{\rho\sigma}(t)\} = \eta_{\alpha\sigma}Q_{\beta\rho}(t) - \eta_{\alpha\rho}Q_{\beta\sigma}(t) + \eta_{\beta\rho}Q_{\alpha\sigma}(t) - \eta_{\beta\sigma}Q_{\alpha\rho}(t) . \tag{10}$$

This completes the demonstration that our Lorentz preserving fields do indeed provide a representation of the Lorentz Lie algebra.

4. Discussion

The special field configurations that we have obtained have aspects of intrinsic interest in cosmological scenarios. They may provide natural seeds for the growth of inhomogeneities in the early Universe [11]. An additional advantage is that these fields allow the use of standard Lorentz invariant formalism even in situations where Lorentz symmetry is violated.

References

[1] Carroll S M, Field G B and Jackiw R 1990 *Phys.Rev.* **D41** 1231
[2] Latorre J I, Pascual P and Tarrach R 1995 *Nuclear Physics B* **437** 60–82 URL `http://www.sciencedirect.com/science/article/pii/0550321394004906`
[3] Colladay D and Kostelecky V A 1998 *Phys.Rev.* **D58** 116002 (*Preprint* `hep-ph/9809521`)
[4] Coleman S R and Glashow S L 1999 *Phys.Rev.* **D59** 116008 (*Preprint* `hep-ph/9812418`)
[5] Colladay D and Kostelecky V A 2001 *Phys. Lett.* **B511** 209–217 (*Preprint* `hep-ph/0104300`)
[6] Myers R C and Pospelov M 2003 *Phys.Rev.Lett.* **90** 211601 (*Preprint* `hep-ph/0301124`)
[7] Jacobson T, Liberati S and Mattingly D 2006 *Annals Phys.* **321** 150–196 (*Preprint* `astro-ph/0505267`)
[8] Mattingly D 2005 *Living Rev.Rel.* **8** 5 (*Preprint* `gr-qc/0502097`)
[9] Maccione L, Taylor A M, Mattingly D M and Liberati S 2009 *JCAP* **0904** 022 (*Preprint* `0902.1756`)
[10] Zwanziger D 1971 *Phys.Rev.* **D3** 880
[11] Ganguly O, Gangopadhyay D and Majumdar P 2011 *Europhys.Lett.* **96** 61001 (*Preprint* `1011.1206`)
[12] Podolsky B and Schwed P 1948 *Rev. Mod. Phys.* **20** 40–50
[13] Ostrogradskii M V 1961 *Complete Collected Works* vol 2 (SSR, Kiev: Akad. Nauk Ukrain) in Russian
[14] Barcelos-Neto J and Braga N R 1989 *Acta Phys.Polon.* **B20** 205
[15] Simon J Z 1990 *Phys.Rev.* **D41** 3720
[16] Morozov A 2008 *Theor.Math.Phys.* **157** 1542–1549 (*Preprint* `0712.0946`)
[17] de Urries F and Julve J 1998 *J.Phys.A* **A31** 6949–6964 (*Preprint* `hep-th/9802115`)
[18] Ganguly O *manuscript in preparation*

Linear higher derivative field theories, entanglement entropy and quantum phase transition

Suman Ghosh and S Shankaranarayanan

School of Physics, Indian Institute of Science Education and Research, Thiruvananthapuram 695016, India

E-mail: suman.ghosh@iisertvm.ac.in

Abstract. Quantum systems described by linear, higher order spatial derivatives of the field exhibit a new kind of phase transition as the coupling strength of the higher order terms varies. This essentially happens because a fundamental change in the ratio of ground state energy eigenvalues, between linear and non-linear regime, takes place which leads to excitation of higher modes. The new phase has inverse relation with area.

Introduction: Entanglement entropy plays crucial role in understanding quantum behavior of macroscopic (black-holes) and microscopic systems [1, 2]. The so-called area law [3, 4, 5, 2] raises the possibility to interpret the Bekenstein-Hawking entropy as the entanglement entropy. Entanglement is also an important factor in understanding quantum phase transitions [6, 7]. It is an useful measure to quantify various aspects of the transitions.

In general, theory of phase-transitions assumes that free energy contains non-linear self-interactions of the order parameter: $F(\phi) = a_2\phi^2 + a_4\phi^4 + \cdots$ [8]. One can generalize the above model by including higher spatial derivatives of the order-parameter [9]. Setting the self-interactions terms to zero ($a_2 = a_4 = 0$), the Hamiltonian for a real scalar field (ϕ) is given by [10, 11, 12, 13]:

$$H = \frac{1}{2}\int d^3x \left[\pi^2(x) + \sum_{m=1}^{\infty} b_m \left(\mathcal{D}^m\phi\right)\left(\mathcal{D}_m\phi\right)\right] , \tag{1}$$

the entanglement entropy shows a sudden discontinuity due to the presence of the term corresponding to $m = 3$. The operator $\mathcal{D}^m$ in the above Hamiltonian is

$$\mathcal{D}_m \equiv \nabla_1\nabla_2 \ldots \nabla_m , \tag{2}$$

where $b_m \equiv \pm\kappa^{-2(m-1)}$ are the appropriate coupling coefficients and κ is the frequency across which the dispersion relation changes from linear to non-linear or vice-versa. To extract the physics, we will consider the following dispersion relation:

$$\omega^2 = k^2 + \frac{\epsilon}{\kappa^2}k^4 + \frac{\tau}{\kappa^4}k^6 , \tag{3}$$

keeping the terms up to $m = 3$ in Hamiltonian (1), where ϵ, τ are (dimensionless) constants. The results presented below have already been submitted for publication [15].

`Simulation technique and results:` The Discretised Hamiltonian (1) along the radial direction of a spherical lattice with lattice spacing a, such that $r \to r_i = ia$; $r_{i+1} - r_i = a$ is given by:

$$H = \sum_j H_j = \frac{1}{2a} \sum_{i,j}^{N} \delta_{ij} \pi_j^2 + \varphi_j K_{ij} \varphi_i \tag{4}$$

where the interaction between them are contained in the off-diagonal elements of the matrix K_{ij}.

Total system size is denoted by a large but finite N. An intermediate point n is chosen, which represents the boundary surface (the *horizon*) with radius $\mathcal{R}$ $(= an)$, that separates the lattice points between two parts.The lattice spacing a appears as an overall factor in the Hamiltonian (4) and in the coupling factor $P \equiv \frac{1}{a^2 \kappa^2}$. This implies that the results for a particular value of N (say, $N = 300$) can be mapped to larger value of N (say, 3000) for smaller value of a. Note that a should be greater than the cutoff scale $(1/\kappa)$ implies $0 \le P \le 1$.

Entanglement is computed as the von Neumann entropy for the reduced density matrix $\rho(r, r')$ [1, 2]: $S^{(P)} = \text{Tr}\,(\rho \ln \rho)$. We compute the entanglement entropy for the discretized Hamiltonian (4) for different P. The computations are done using Matlab for the lattice size $N = 300, 50 \le n \le 295$ and the relative error in the computation is 10^{-5}. Figures below show the results for the following two different scenarios:

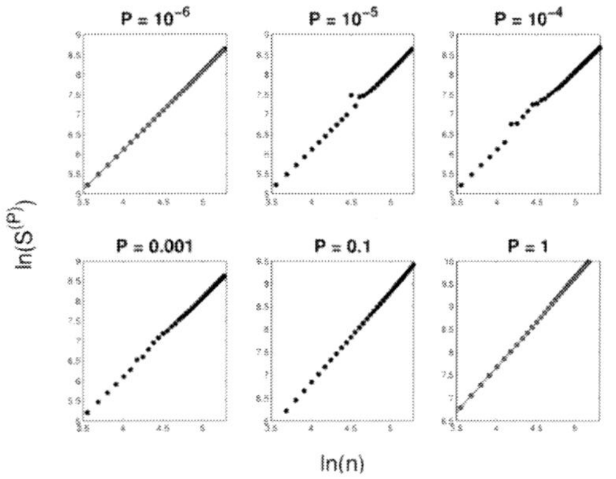

Figure 1. Log-log plot of von Neumann entropy, S, versus the scaled radius of the sphere $\mathcal{R}/a = n$, for $\tau = 0$, $\epsilon = 1$ and different P. The dots represent the numerical output and solid lines denote lines of best fit.

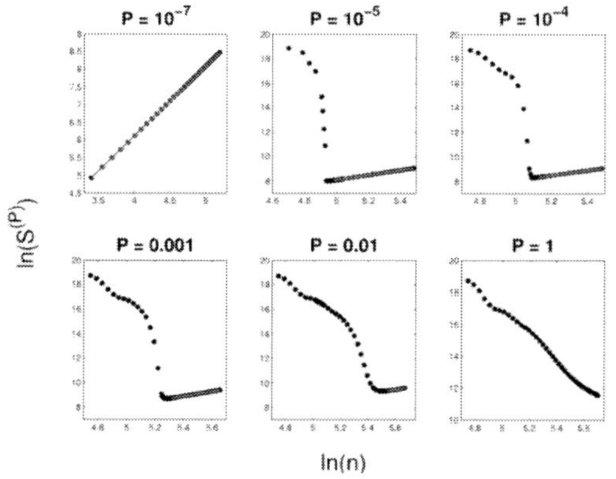

Figure 2. Log-log plot of von Neumann entropy, S, versus the scaled radius of the sphere $\mathcal{R}/a = n$, for $\tau = 1$, $\epsilon = 0$ and different P. n_c for $P = 10^{-5}, 10^{-4}, 10^{-3}$, respectively, are 139, 167 and 197.

(1) $\tau = 0$, $\epsilon = 1$: The best fit (solid) curves in the two asymptotic regimes, $P \to 0$ and $P \to 1$, show that the entropy scales approximately as area $S \sim (\mathcal{R}/a)^2$. With increasing P, however, the prefactor increases. For $P = 10^{-5}$ and 10^{-4}, around the transition region between linear and non-linear, entropy increases by an order.

(2) $\tau = 1$, $\epsilon = 0$: (i) For small P, entropy scales as area. (ii) As we increase P, above a *critical* value of P, say P_c (here $P_c \sim 10^{-6}$) a new phase appears in the log – log plots of entropy vs n. (iii) For $P > P_c$, scaling of $S^{(P)}$ changes across some *critical* value of n, say n_c (e.g. $n_c \sim 137$ for $P = 10^{-5}$). For $n > n_c$, $S^{(P)}$ increases with n and scales approximately to area. However, for $n < n_c$, $S^{(P)}$ increases with decreasing n [1] (iv) With increasing P, n_c increases.

[1] Note that as $n \to 0$, the entropy becomes zero as the number of degrees of freedom gradually vanishes.

The above results indicate that higher derivative term for $m = 3$ modifies the vacuum such that the entropy jumps by few orders of magnitude close to the critical point.

Fig. (3) shows the entropy distribution per partial wave for two values of n — $n = 200$ (which falls in the *area-law* region in Fig. 2) and $n = 137$ (from the transition region in Fig. 2) for $\epsilon = 0, \tau = 1$ and $P = 10^{-5}$. For $n = 200$, the contribution of large ℓ falls off rapidly implying that only low ℓ-modes contribute to the entropy. However, for $n = 137$, higher partial waves contribute significantly to the entropy compared to low ℓ.

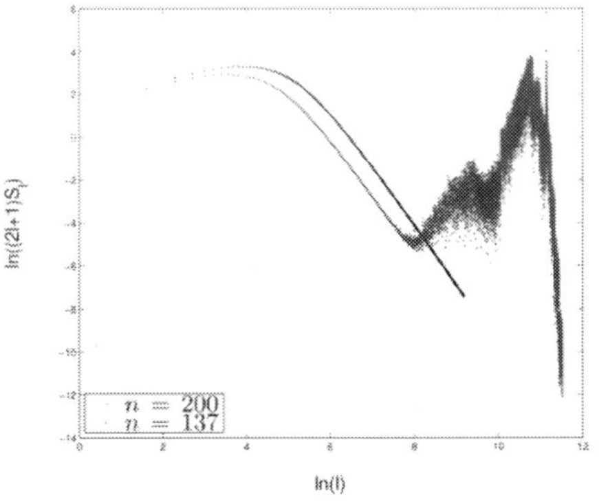

Figure 3. Log-Log plot of the entropy distribution per partial wave $(2\ell + 1)S_\ell$ vs ℓ for $n = 200$ and 137 for $P = 10^{-5}$. For a specific n and P, the asymptotic analysis shows that modes higher than a critical value $\ell_c \sim P^{-1/2}n$ contribute more than the ones lower than ℓ_c. Which also implies that for a fixed mode ℓ_c, this cross-over appear at higher n with increasing P as found in Fig. (2).

`Undestanding the results:` The higher derivative terms dominate with decreasing length and hence, for a fixed P the inverse scaling phase appears below certain n_c. To understand the cause of excitation of high ℓ-modes (for $P > P_c$), let us consider a simple quantum mechanical model of a particle in a box. The Schrödinger equation for a non-relativistic particle in 1-dimensional infinite square well with a general dispersion relation, $F(k) \propto k^m$, is given by

$$\frac{1}{\kappa^{2(m-1)}} \frac{d^{2m}\psi}{dx^{2m}} + E\psi = 0, \tag{5}$$

Eq. (5) can be solved for all m with appropriate boundary conditions at $x = 0$ and $x = L$. Let us consider that this system obeys the linear dispersion relation ($F(k) = k$) at low-energies and non-linear dispersion relation ($F(k) = k^m$) at high energies. Then Eq. (5) implies that the ratio of ground state energy eigenvalues in these two regimes is:

$$\frac{\text{linear dispersion ground state energy}}{\text{non-linear dispersion ground state energy}} = \left[\frac{L\kappa}{\pi}\right]^{2(m-1)} \tag{6}$$

Eq. (6) implies that, for $\kappa < \pi/L$, the ground state energy eigenvalue of the system satisfying linear dispersion relation is lower compared to that of non-linear dispersion relations. As in the field theory model, where there is a cross-over from the linear to non-linear regime, with increasing P (the condition $\kappa < \pi/L$ implies that the cross-over occurs for $P > P_c$), the system needs to readjust in such a way that the ground state energy of the system increases. In other words, the cross-over of the dispersion relation catalyses *larger population* of higher energy quantum states compared to low-energy states.

We also argue that the change in the ground state is related to a *quantum phase transition* based on the following fact. The two point correlation (Wightman) Function $G^+(x^\mu, y^\mu)$ for $m = 1$ and $m = 2$, respectively, are $\frac{1}{4\pi^2 r^2}$ and $\frac{\kappa}{8\pi r}$ where $r = |\vec{x} - \vec{y}|$. Note that in linear dispersion regime the interaction is *localised* which explains the area-law behavior of the entropy. When the second order correction dominates, correlation *decays* slowly (w.r.t. the linear dispersion scenario) with increasing distance which is approximately same as having nearest neighbour interaction. This implies that including the 3rd order derivative term will further increase the correlation length.

`Discussions:` We have shown that linear, higher spatial derivative theories give rise to a new kind of quantum phase transition. The new phase, with inverse scaling, appears due to excitation of higher modes which can be understood by considering a toy model of particle in a box where a cross-over beween linear and non-linear dispersion happens.

As the size of a black-hole decreases, due to Hawking radiation, the curvature of the event-horizon increases and hence one needs to include higher curvature terms [14, 16, 17, 18]. Our analysis shows that as the size of the black-hole decreases, below a critical radius, the scaling of entropy changes from area to an inverse-scaling wrt size of the horizon.

Currently we are investigating how n_c/N varies with $N \to \infty$ (the thermodynamic limit). We hope to report on more detailed understanding of this phenomenon elsewhere [15].

Acknowledgments

This work is supported by the DST, Government of India through Ramanujan fellowship and Max Planck-India Partner Group on Gravity and Cosmology.

References

[1] Horodecki R, Horodecki P, Horodecki M and Horodecki K 2009 *Rev. Mod. Phys.* **81** 865
[2] Eisert J, Cramer M and Plenio M B 2010 *Rev. Mod. Phys.* **82**, 277
[3] Bombelli L, Koul R K, Lee J H and Sorkin R D 1986 *Phys. Rev.* D **34** 373
[4] Srednicki M 1993 *Phys. Rev. Lett.* **71** 666
[5] Das S, Shankaranarayanan S and Sur S 2008 *Phys. Rev.* D **77** 064013
[6] Sachdev S 2001 *Quantum phase transitions* (Cambridge University Press)
[7] Carr L D 2011 *Understanding quantum phase transitions* (CRC Press)
[8] Toledano J C and Toledano P 1987 *The Landau Theory of Phase Transitions. Application to Structural, Incommensurate, Magnetic and Liquid Crystal Systems* (World Scientific)
[9] Honreich R M, Luban M and Shritman S 1975 *Phys. Rev. Lett.* **35** 1678
[10] Unruh W G 1995 *Phys. Rev.* D **51** 2827
[11] Corley S and Jacobson T 1996 *Phys. Rev.* D **54** 1568
[12] Padmanabhan T 1999 *Phys. Rev.* D **59** 124012
[13] Visser M 2009 *Phys. Rev.* D **80** 025011
[14] Wald R 1993 *Phys. Rev.* D **48** R3427
[15] Ghosh S and Shankaranarayanan S 2012 Scaling of entanglement in higher derivative theories and quantum phase transition *submitted to the journal*
[16] Jacobson J and Myers R C 1993 *Phys. Rev. Lett.* **70** 3684
[17] Sen A 2008 *Gen. Rel. Grav.* **40** 2249
[18] Kothawala D and Padmanabhan T 2009 *Phys. Rev.* D bf 79 104020

Journal of Physics: Conference Series **405** (2012) 012017 doi:10.1088/1742-6596/405/1/012017

Geodesic Motions near a Five-dimensional Reissner-Nordström Anti-de Sitter black hole

Sarbari Guha[1] **and Pinaki Bhattacharya**[2]

[1] Department of Physics, St. Xavier's College (Autonomous), Kolkata 700016, India
[2] Gopal Nagar High School, Singur 712409, West Bengal, India

E-mail: srayguha@yahoo.com

Abstract. We have studied the geodesics of neutral particles near a non-rotating, charged five-dimensional Reissner-Nordström Anti-de Sitter black hole using the effective potential analysis and the dynamical systems analysis. The effective potential analysis is used to determine the location of the horizon and to study radial and circular trajectories. The dynamical systems method is used to determine the stability and the fixed points of the phase trajectories.

1. Introduction

With the development of string theory, the study of black holes in higher-dimensional spacetimes [1] have gained momentum, especially in the first decade of this millennium [2]. Static, spherically symmetric exterior vacuum solutions of the braneworld models were first proposed by Dadhich and others [3]. Extensive studies in higher-dimensional spacetimes over the last few decades have led many authors to investigate the geodesic motions in such spacetimes [4]. Motion of massive particles around a rotating black hole in a braneworld has been studied [5] and the effective potentials for radial null geodesics in RN-dS and Kerr-dS spacetimes were analyzed [6]. Analytic solutions of the geodesic equations in Schwarzschild-(Anti-)de Sitter spacetimes [7] and the motion of massive particles in 4-dimensions and in higher dimensions, has been analysed. Here we have investigated the radial and circular trajectories for photons and massive particles in a five-dimensional RN-AdS spacetime and have determined the fixed points of the phase trajectories. For such a non-rotating charged black hole in Anti-de Sitter spacetime, the solutions are uniquely characterized by their mass, charge and the cosmological constant [8].

2. Preliminaries

We consider a 5D spacetime with negative cosmological constant Λ. The radius of curvature $l = \sqrt{-\frac{3}{\Lambda}}$ of the spacetime provides the length scale necessary to have a horizon. The exterior metric of the black hole field is given by

$$dS^2 = -f(r)dt^2 + \frac{dr^2}{f(r)} + r^2 d\Omega_3^2 = -\frac{\triangle}{r^4}dt^2 + \frac{r^4}{\triangle}dr^2 + r^2 d\Omega_3^2. \quad (1)$$

The Lapse function is defined as: $f(r) = 1 - \left(\frac{2M}{r}\right)^2 + \left(\frac{q^2}{r^2}\right)^2 - \frac{\Lambda r^2}{6}$.

Published under licence by IOP Publishing Ltd

For a given M, q and Λ, the ***horizon function*** $\triangle$ depends only on r. The nature of the intrinsic singularity at $r=0$, depends on Λ and q, which we choose such that the spacetime do not have any spacelike naked singularity. The lapse function and the effective potential vanishes at the real, positive zeros of the horizon function $\triangle$, indicating the location of the horizons. The variation of the effective potential V_{eff} with specific radius r/M in the case of radial motion of massive particles, in the field of the black hole, are shown in Fig.1 and Fig. 2 for different range of values of the r/M. The location of the horizons can be easily identified from these figures.

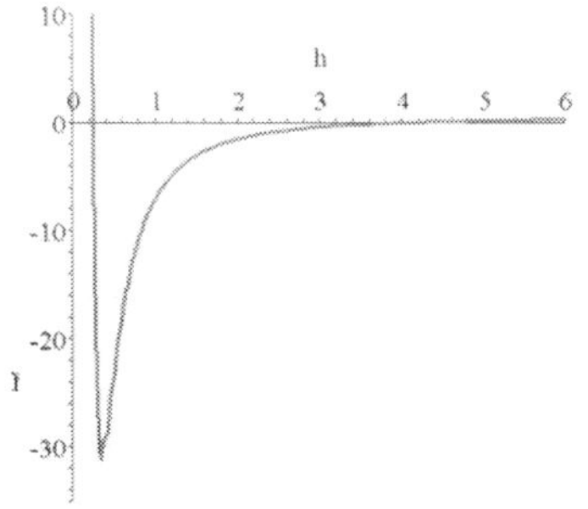

Figure 1. Plot of effective potential V_{eff} vs radial distance $h = r/M$ in the case of radial motion of massive particles with $q/M = 0.5$ and $\Lambda = -0.0005$ for $r/M = 0$ to 6.

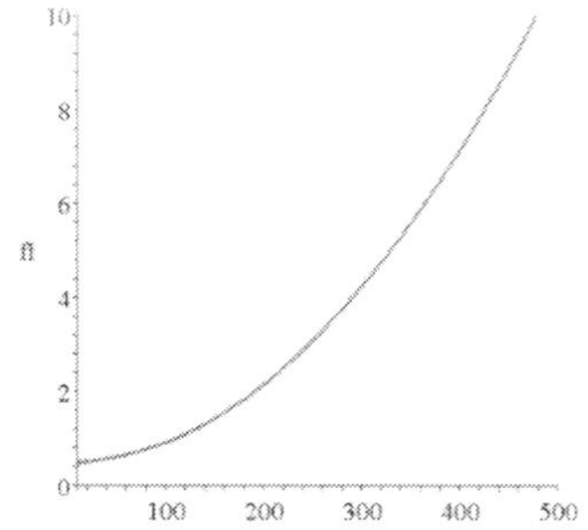

Figure 2. Plot of effective potential V_{eff} vs radial distance $h = r/M$ in the case of radial motion of massive particles with $q/M = 0.5$ and $\Lambda = -0.0005$ for $r/M = 6$ to 500.

3. Five-dimensional Geodesics

Due to **spherical symmetry**, we analyze the motion of neutral particles on the **equatorial hyperplane**, $\theta, \phi = \pi/2$. We have the following equations:

$$\frac{d^2t}{d\lambda^2} + \frac{B(r)}{A(r)}\frac{dt}{d\lambda}\frac{dr}{d\lambda} = 0, \qquad \frac{d^2\psi}{d\lambda^2} + \frac{1}{r}\frac{dr}{d\lambda}\frac{d\psi}{d\lambda} = 0, \tag{2}$$

$$\frac{d^2r}{d\lambda^2} + A(r)B(r)\left(\frac{dt}{d\lambda}\right)^2 - \frac{B(r)}{A(r)}\left(\frac{dr}{d\lambda}\right)^2 + rA(r)\left(\frac{d\psi}{d\lambda}\right)^2 = 0, \tag{3}$$

where $A(r) = -f(r)$ and $B(r) = \frac{1}{r}\left(-\left(\frac{2M}{r}\right)^2 + 2\left(\frac{q^2}{r^2}\right)^2 + \frac{\Lambda r^2}{6}\right)$.

4. Effective Potential Analysis

The **Lagrangian** for particle motion is: $Ł = -\left(1 - \left(\frac{2M}{r}\right)^2 + \left(\frac{q^2}{r^2}\right)^2 - \frac{\Lambda r^2}{6}\right)\dot{t}^2$

$$+\left(\frac{\dot{r}^2}{1-\left(\frac{2M}{r}\right)^2+\left(\frac{q^2}{r^2}\right)^2-\frac{\Lambda r^2}{6}} + r^2(\dot{\theta}^2 + sin^2\theta\dot{\phi}^2 + sin^2\theta sin^2\phi\dot{\psi}^2)\right).$$

There are two conserved quantities: **Energy**, $E = g_{tt}\frac{dt}{d\lambda} = -f(r)\frac{dt}{d\lambda} = A(r)\frac{dt}{d\lambda}$, and **momentum** $L = r^2\frac{d\psi}{d\lambda}$ conjugate to ψ.

On the equatorial hyperplane we have, $\left(\frac{dr}{d\lambda}\right)^2 = E^2 - \frac{\triangle}{r^4}\left(\epsilon + \frac{L^2}{r^2}\right) = E^2 + A(r)\left(\epsilon + \frac{L^2}{r^2}\right)$

so that $\frac{1}{2}\left(\frac{dr}{d\lambda}\right)^2 = E_{eff} - V_{eff}(r)$, where $E_{eff} = \frac{1}{2}E^2$ and
$V_{eff}(r) = \frac{\triangle}{2r^4}\left(\epsilon + \frac{L^2}{r^2}\right) = \frac{1}{2}\left(1 - \left(\frac{2M}{r}\right)^2 + \left(\frac{q^2}{r^2}\right)^2 - \frac{\Lambda r^2}{6}\right)\left(\epsilon + \frac{L^2}{r^2}\right)$.

Radial motion: Here $L = 0$. For bound states of massive particles, we have $\frac{4M^2}{r^3} > \frac{2q^4}{r^5} + \frac{\Lambda r}{6}$.

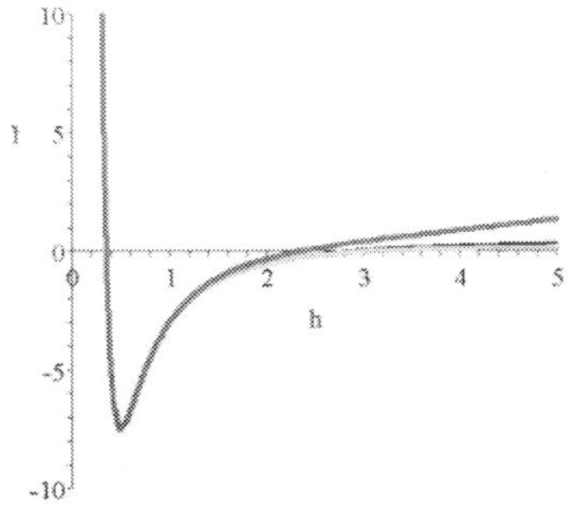

Figure 3. V_{eff} vs radial distance $h = r/M$ in the case of radial motion of massive particles with $q/M = 1/\sqrt{2}$ and $\Lambda = -0.5, -0.0005$ (orange and black).

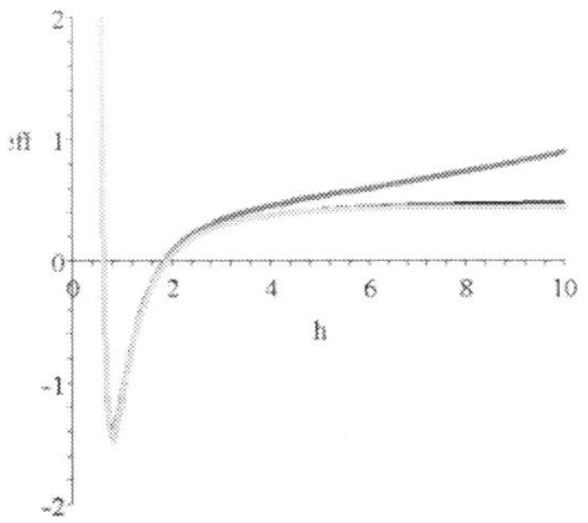

Figure 4. V_{eff} vs radial distance for circular motion of massive particles with $q/M = 1.1$ and $\Lambda = -0.05, -0.0005$ (orange and black).

Circular motion: We now introduce the variable change $u = r^{-1}$. For equilibrium circular orbits,we have
$L^2 = \frac{\epsilon[6q^4u^6 - 12M^2u^4 + \Lambda/2]}{3u^4[-3q^4u^4 + 8M^2u^2 - 1]}$ and $E^2 = \frac{\epsilon[6q^4u^6 - 24M^2u^4 + 6u^2 - \Lambda][3q^4u^6 - 12M^2u^4 + 3u^2 - \Lambda]}{18u^4[3q^4u^4 - 8M^2u^2 + 1]}$.

For photons, we have $u^2 = \frac{4M^2 \pm \sqrt{16M^4 - 3q^4}}{3q^4}$. Therefore, circular orbits occur for only two values of u in the case of photons.

4.1. Stability of the orbits

The stable circular orbits occur for those values of r which are located at the local minimum of the potential. Here we obtain

$$V_{eff}(r) = \frac{1}{2}\left(1 - \left(\frac{2M}{r}\right)^2 + \left(\frac{q^2}{r^2}\right)^2 - \frac{\Lambda r^2}{6}\right)\left(1 + \frac{L^2}{r^2}\right). \quad (4)$$

For $4M^2L^2 = q^4$ and $4M^2 \neq L^2$, we have $r^2_{ISCOM} = \frac{\sqrt{30}Lq^2}{(4M^2 - L^2) + \sqrt{(4M^2 - L^2)^2 + 30L^2q^4\Lambda}}$, giving only two real values of r for **massive particles**, if $(4M^2 - L^2)^2 > |30L^2q^4\Lambda|$.

The minimum radius of stable circular orbits of **photons** is obtained as:

$$r_{mc} > \frac{\sqrt{15}q^2}{\sqrt{12M^2 - \sqrt{144M^4 - 15q^4}}}.$$

5. Dynamical Systems Analysis

Let us define **three new variables** [9]: $U = \frac{dt}{d\lambda}$, $V = \frac{dr}{d\lambda}$ and $W = \frac{d\psi}{d\lambda}$.

Thus the geodesics equations are:

$\frac{dU}{d\lambda} + \frac{B(r)}{A(r)}UV = 0$, $\frac{dW}{d\lambda} + \frac{1}{r}VW = 0$ and $\frac{dV}{d\lambda} + A(r)B(r)U^2 - \frac{B(r)}{A(r)}V^2 + rA(r)W^2 = 0$, which are related through $A(r)U^2 - \frac{1}{A(r)}V^2 + r^2W^2 = -\epsilon$.

Real non-linear dynamical system:

We now assume $\frac{dU}{d\lambda} = H(U,V,r)$, $\frac{dr}{d\lambda} = V$ and $\frac{dV}{d\lambda} = J(U,V,r)$, where $H(U,V,r) = -\frac{B(r)}{A(r)}UV$, $J(U,V,r) = \frac{\epsilon A(r)}{r} - \frac{A(r)}{r}(rB(r) - A(r))U^2 - \frac{1}{rA(r)}(A(r) - rB(r))V^2$.

Fixed Points: For a black hole with a given M and q, (0,0) is the unique fixed point on the (U-V) plane for null geodesics. The timelike geodesics possess definite fixed points.

The **phase evolution** of the system on the (U-V) phase plane is determined from the condition $\frac{dV}{dU} = \frac{J}{H}$, provided $H \neq 0$ (except at the point (0,0) for null geodesics). Simplifying, we get $f_1U^2 - f_2V^2 - f_3 = 0$, where f_1, f_2 and f_3 are functions of r. For null geodesics, we get $f_3 = 0$.

Thus the geodesics of massive particles near a black hole in a RN-AdS_5 possesses definite fixed points and the orbits are either elliptic (periodic bound) or hyperbolic (escape orbits). Moreover, for a black hole of a given charge and mass, the null geodesics possesses a unique fixed point ($U_0 = 0$, $V_0 = 0$, r_0). The trajectories are linear, with their slopes changing according to the values of the parameters f_1 and f_2. Hence the null geodesics are terminating orbits.

6. Conclusions

We have determined the location of the black hole horizons from the plot of effective potential. The nature of the trajectories depend upon the particle energies and their angular momenta, as well as on Λ and q. The radius of the innermost stable circular orbit of massive particles is totally defined in terms of their angular momenta and M and q of the black hole. Photons trace out circular trajectories for only two distinct values of r. The geodesics of massive particles near a black hole in a RN-AdS_5 are either periodic bound or escape orbits and these have definite fixed points. The null geodesics have a unique fixed point and these are terminating orbits.

[1] Emparan R and Reall H S 2008 *Liv. Rev. Rel.* 2008-6 (http://www.livingreviews.org/lrr-2008-6)
[2] Gibbons G W, Lu H, Page D N and Pope C N 2004 *Phys. Rev. Lett.* **93** 171102; Sen A 2005 *J. High Energy Phys.* 07(2005)073; Frolov V and Stojkovic D 2003 *Phys. Rev. D* **68** 064011
[3] Dadhich N K, Maartens R, Papodopoulos P and Rezania V 2000 *Phys. Lett. B* **487** 1
[4] Page D N, Kubiznak D, Vasudevan M and Krtous P 2007 *Phys. Rev. Lett.* **98** 061102; Cardoso V, Cavaglia M and Gualtieri L 2006 *Phys. Rev. Lett.* **96** 071301; Konoplya R A and Zhidenko A 2008 *Phys. Rev. D* **78** 104017; Cruz N, Olivares M and Villanueva J R 2005 *Class. Quant. Grav.* **22** 1167
[5] Abdujabbarov A and Ahmedov B 2010 *Phys. Rev. D* **81** 044022
[6] Stuchlik Z and Calvani M 1991 *Gen. Relativ. Grav.* **23** 507
[7] Hackmann E and Lämmerzahl C 2008 *Phys. Rev. Lett.* **100** 171101
[8] Gibbons G W, Ida D and Shiromizu T 2002 *Prog. Theor. Phys. Suppl.* **148** 284
[9] Guha S and Chakraborty S 2010 *Gen. Relativ. Grav.* **42** 1739
[10] Wainwright J and Ellis G F R 1997 *Dynamical Systems in Cosmology* (Cambridge: Cambridge University Press)
[11] Dahia F, Romero C, da Silva L F P and Tavakol R 2007 *J. Math. Phys.* **48** 072501, Dahia F, Romero C, da Silva L F P and Tavakol R 2008 *Gen. Rel. Grav.* **40** 1341

Wide binaries as a critical test for Gravity theories

X Hernandez, M A. Jiménez and C Allen

Instituto de Astronomía, Universidad Nacional Autónoma de México, AP 70-264, Distrito Federal 04510, México

E-mail: xavier@astro.una,.mx

Abstract. Assuming Newton's gravity and GR to be valid at all scales leads to the dark matter hypothesis as a requirement demanded by the observed dynamics and measured baryonic content at galactic and extragalactic scales. Alternatively, modified gravity scenarios where a change of regime appears at acceleration scales $a < a_0$ have been proposed. This modified regime at $a < a_0$ will generically be characterised by equilibrium velocities which become independent of distance. Here we identify a critical test in this debate and we propose its application to samples of wide binary stars. Since for $1M_\odot$ systems the acceleration drops below a_0 at scales of around 7000 AU, a statistical survey of wide binaries with relative velocities and separations reaching 10^4 AU and beyond should prove useful to the above debate. We apply the proposed test to the best currently available data. Results show a constant upper limit to the relative velocities in wide binaries which is independent of separation for over three orders of magnitude, in analogy with galactic flat rotation curves in the same $a < a_0$ acceleration regime. Our results are suggestive of a breakdown of Kepler's third law beyond $a \approx a_0$ scales, in accordance with generic predictions of modified gravity theories designed not to require any dark matter at galactic scales and beyond.

1. Introduction

Over the past years the dominant explanation for the large mass to light ratios inferred for galactic and meta-galactic systems, that these are embedded within massive dark matter halos, has begun to be challenged. Direct detection of the dark matter particles, in spite of decades of extensive and dedicated searches, remains lacking. Numerous alternative theories of gravity have appeared (e.g. TeVeS of [1], F(R) theories e.g. [8]) mostly grounded on geometrical extensions to General Relativity, and leading to laws of gravity which in the low acceleration regime, mimic the MOdified Newtonian Dynamics (MOND) fitting formulas e.g. [7]. Similarly, [6] have explored MOND not as a modification to Newton's second law, but as a modified gravitational force law in the Newtonian regime, finding a good agreement with observed dynamics across galactic scales without requiring dark matter. Recently [2] have constructed an $f(R)$ extension to general relativity which in the low velocity limit converges to the above approach.

A generic feature of all such modified gravity schemes is the appearance of an acceleration scale, a_0, above which classical gravity is recovered, and below which the dark matter mimicking regime appears. This results in a general prediction; all systems where $a >> a_0$ should appear as devoid of dark matter, and all systems where $a << a_0$ should appear as dark matter dominated. It is interesting that no $a >> a_0$ system has ever been detected where dark matter needs to be invoked. On the other hand, the latter condition furnishes a testable prediction in relation to the orbits of wide binaries. For test particles in orbit around a $1M_\odot$ star, in circular orbits

Journal of Physics: Conference Series **405** (2012) 012018 doi:10.1088/1742-6596/405/1/012018

of radius s, the acceleration is expected to drop below $a_0 \approx 1.2 \times 10^{-10} m/s^2$ for s >7000 AU= $3.4 \times 10^{-2} pc$. This provides a test for the dark matter/modified theories of gravity debate; the relative velocities of components of binary stars with large physical separations should deviate from Kepler's third law. Seen as an equivalent Newtonian force law, beyond $s \approx$7000 AU the gravitational force should gradually switch from the classical form of $F_N = GM/s^2$ to $F_{MG} = (GMa_0)^{1/2}/s$, and the orbital velocity, $V^2/s = F$, should no longer decrease with separation, but settle at a constant value, dependent only on the total mass of the system through $V = (GMa_0)^{1/4}$. Under modified gravity theories, binary stars with separations beyond 7000 AU should exhibit "flat rotation curves" and a "Tully-Fisher relation", as galactic systems in the same acceleration regime do.

We apply this test to the binaries of two recent catalogues containing relative velocities and separations. The two catalogues are entirely independent in their approaches. The first one, [8], comprises 280 systems from the *Hipparcos* satellite, while the second, [3], identifies 1,250 ones from the Sloan Digital Sky Survey (SDSS). Details can be found in [4].

2. Relative velocity distributions for wide binaries under Newtonian expectations

The Newtonian prediction for the relative velocities of the two components of binaries having circular orbits, when plotted against the binary physical separation, s, is for a scaling of $\Delta V \propto s^{-1/2}$, essentially following Kepler's third law, provided the range of masses involved were narrow. In a relative proper motion sample however, only two components of the relative velocity appear, as velocity along the line of sight to the binary leads to no proper motion. Thus, orbital projection plays a part, with systems having orbital planes along the line of sight sometimes appearing as having no relative proper motions. A further effect comes from any degree of orbital ellipticity present; it is hence clear that the trend for $\Delta V \propto s^{-1/2}$ described above, will only provide an upper limit to the distribution of projected ΔV vs. s expected in any real observed sample, even if only a narrow range of masses is included. One should expect a range of measured values of projected ΔV at a fixed observed projected s, all extending below the Newtonian limit, which for equal mass binaries in circular orbits gives $\Delta V_N = 2\left(\frac{GM}{s}\right)^{1/2}$.

Further, over time, the orbital parameters of binaries will evolve due to the effects of Galactic tidal forces and dynamical encounters with other stars in the field, specially in the case of wide binaries. To first order, one would expect little evolution for binaries tighter than the tidal limit of 1.7pc, and the eventual dissolution of wider systems. A very detailed study of all these points has recently appeared, [5]. These authors numerically follow populations of 50,000 $1M_\odot$ binaries in the Galactic environment, accounting for the evolution of the orbital parameters of each due to the cumulative effects of the Galactic tidal field at the Solar radius. Also, the effects of close and long range encounters with other stars in the field are carefully included, to yield a present day distribution of separations and relative velocities for an extensive population of wide binaries, under Newtonian Gravity.

It is found that when many wide binaries cross their Jacobi radius, the two components remain fairly close by in both coordinate and velocity space. Thus, in any real wide binary search a number of wide pairs with separations larger than their Jacobi radii will appear. Finally, [5] obtain the RMS one-dimensional relative velocity difference, ΔV_{1D}, projected along a line of sight, for the entire populations of binaries dynamically evolved over 10 Gyr to today, as plotted against the projected separation on the sky for each pair. The expected Keplerian fall of $\Delta V_{1D} \propto s^{-1/2}$ for separations below 1.7 pc is obtained, followed by a slight rise in ΔV_{1D} as wide systems cross the Jacobi radius threshold. ΔV_{1D} then settles at RMS values of $\approx 0.1 km/s$. This represents the best currently available estimate of how relative velocities should scale with projected separations for binary stars (both bound and in the process of dissolving in the Galactic tides) under Newtonian gravity. We see that all we need is a large sample of relative proper motion and binary separation measurements to test the Newtonian prediction for the RMS values of the 1 dimensional relative velocities of [5], and to contrast the $\Delta V_N \propto s^{-1/2}$ and the $\Delta V_{MG} = cte.$ predictions for the upper envelope of the ΔV vs. s distributions.

3. Observed wide binary Samples

In the [8] catalogue wide binaries are identified by assigning a probability above chance alignment to the systems by carefully comparing to the underlying background (and its variations) in a 5 dimensional parameter space of proper motions and spatial positions. We keep only binaries with a probability of non-chance alignment greater than 0.9. The binary search criteria used by the authors requires that the proposed binary should have no near neighbours; the projected separation between the two components is thus always many times smaller than the typical interstellar separation. We use the reported distances to the primaries, where errors are smallest, to calculate projected ΔV and projected s from the measured $\Delta\mu$ and $\Delta\theta$ values reported. Although the use of *Hipparcos* measurements guarantees the best available quality in the data, we have also removed all binaries for which the final signal to noise ratio in the relative velocities was lower than 0.3.

We are left with a sample of 280 binaries, having distances to the Sun within $6 < d < 100$ in pc. The data show a perfectly flat upper envelope in a ΔV vs. projected s, [4]. The average signal to noise ratio for the data is 1.7, with an average error on ΔV of 0.83 km/s, which considering a 2σ factor from the top of the distribution to the real underlying upper limit for the sample, results in 3 km/s as our estimate of the actual physical upper limit in ΔV.

The Sloan low mass wide pairs catalogue (SLoWPoKES) of [3] contains a little over 1,200 wide binaries with relative proper motions for each pair, distances and angular separations. Also, extreme care was taken to include only physical binaries, with a full galactic population model used to exclude chance alignment stars using galactic coordinates and galactic velocities, resulting in an estimate of fewer than 2% of false positives. This yields only isolated binaries with no neighbours within many times the internal binary separation. Again, we use the reported distances to the primaries to calculate projected ΔV and projected s from the measured $\Delta\mu$, $\Delta\theta$ and d values reported by [3], to obtain a sample of 417 binaries.

The upper envelope of the distribution of ΔV from this catalogue does not comply with Kepler's third law. As was the case with the *Hipparcos* sample, the upper envelope describes a flat line, as expected under modified gravity schemes. The average signal to noise in ΔV for the [3] catalogue is 0.48, with an average error on ΔV of 12 km/s, which considering a 2σ factor from the top of the complete distribution to the real underlying upper limit gives the same $3km/s$ as obtained for the [8] *Hipparcos* catalogue.

Figure (1) shows the RMS value of the one-dimensional relative velocity difference for both of the samples discussed. The error bars give the error propagation on $\Delta\mu$ and d. We construct ΔV_{1D} by considering only one coordinate of the two available from the relative motion on the plane of the sky. Thus, each binary can furnish two ΔV_{1D} measurements, which statistically should not introduce any bias. Indeed, using only $\Delta\mu_l$ or only $\Delta\mu_b$ or both for each binary, yields the same mean values for the points shown. The small solid error bars result from considering an enlarged sample where each binary contributes two ΔV_{1D} measurements, while the larger dotted ones come from considering each binary only once, and do not change if we consider only $\Delta\mu_l$ or only $\Delta\mu_b$. The series of small $log(s)$ interval data are for the *Hipparcos* catalogue of [8], while the two broader crosses show results for the [3] SDSS sample.

The solid curve is the Newtonian prediction of the full Galactic evolutionary model of [5] for binaries, both bound and in the process of dissolving. Note that the results of this simulation deviate from Kepler's law for s larger than the Newtonian Jacobi radius of $1.7pc$, whereas the discrepancy with the observed samples also occurs at much smaller separations. Even considering the large error bars, where each binary contributes only one ΔV_{1D} value, we see eight points lying beyond 1σ, making the probability of consistency between this prediction and the observations of less than $(0.272)^8 = 3 \times 10^{-5}$.

We obtain a constant RMS value for ΔV_{1D} of 1 km/s, in qualitative agreement with expectations from modified gravity schemes. The vertical line marks $a = a_0$; we see the data departing from the Newtonian prediction outwards of this line, and not before. The two independent catalogues, each using different sets of selection criteria, each perhaps subject to its own independent systematics, are consistent with the same result, a constant horizontal upper envelope for the distribution of relative velocities on the plane of the sky at an intrinsic value of 3 $km/s \pm 1$ km/s, extending over 3 orders of magnitude in s, with a constant RMS ΔV_{1D} value consistent with 1 $km/s \pm 0.5$ km/s. This supports the interpretation of the effect detected as the generic prediction of modified gravity theories.

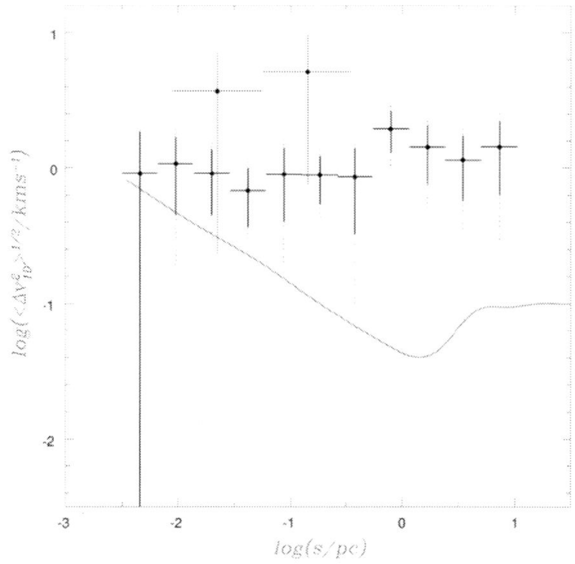

Figure 1. The solid curve gives the RMS values for one dimensional projected relative velocities as a function of projected separations, for the dynamical modelling of large populations of wide binaries evolving in the Galactic environment, taken from [5]. The same quantity for the data from the catalogues analysed is given by the points with error bars; those with narrow $log(s)$ intervals being from the *Hipparcos* sample of [8], and those two with wide $log(s)$ intervals coming from the SDSS sample of [3]

4. Conclusions

We identify a critical test in the classical gravity/modified gravity debate, using the relative velocities of wide binaries with separations in excess of 7000 AU, as these occupy the $a < a_0$ regime. We present a first application of this test using the best currently available data. Results show constant relative RMS velocities for the binary stars in question. This is quantitatively inconsistent with detailed predictions of Newtonian dynamical models for binaries evolving in the local galactic environment. Our results are qualitatively in accordance with generic modified gravity models which explain galactic dynamics in the absence of dark matter.

References

[1] Bekenstein, J., 2004, Phys. Rev. D., 70, 3509
[2] Bernal T., Capozziello S., Hidalgo J. C., Mendoza S., 2011, Eur. Phys. J. C, 71, 1794
[3] Dhital, S., West, A. A., Stassun, K. G., Bochanski, J. J., 2010, AJ, 139, 2566
[4] Hernandez, X., Jimnez, M. A., Allen, C., 2012, EPJC, 72, 1884
[5] Jiang, Y. F., & Tremaine, S., 2010, MNRAS, 401, 977
[6] Mendoza, S., Hernandez, X., Hidalgo, J. C., Bernal, T., 2011, MNRAS, 411, 226
[7] Milgrom, M., 2010, Phys. Rev. D., 82, 043523
[8] Shaya, E. J., Olling, R. P., 2011, ApJS, 192, 2

International Conference on Modern Perspectives of Cosmology and Gravitation (COSGRAV12) IOP Publishing
Journal of Physics: Conference Series **405** (2012) 012019 doi:10.1088/1742-6596/405/1/012019

Gravitational waves from $f(R)$ theory and its detection using spherical antenna

P Prasia and V C Kuriakose
Department of Physics, Cochin University of Science and Technology, Kochi 682022, Kerala, India
E-mail: prasiapankunni@cusat.ac.in, vck@cusat.ac.in

Abstract. Gravitational waves generated from a metric $f(R)$ theory of the form $f(R) = R + \lambda R^2$ are studied. Two solutions for the metric perturbation, one corresponding to massless mode as in General Theory of Relativity and the other a massive mode, are obtained. The massive mode is found to follow non-null geodesics indicating that the gravitational wave arising from such a mode travels with a speed less than that of light. The detection of the massive mode of the gravitational waves using spherical antenna is studied. The response of the antenna to massive modes show better energy sensitivity and high directionality compared to the ordinary gravitational waves from General Theory of Relativity.

1. Introduction

The fact that General Theory of Relativity(GTR) is not renormalizable and the current scenario of the accelerating universe call for searching new theories of gravitation. $f(R)$ theories are one such attempt[**1**]. $f(R)$ theories come out as a straightforward generalization of the Einstein-Hilbert action of GTR and can be written as,

$$S = \frac{1}{2\kappa}\int d^4x\sqrt{-g}f(R), \qquad (1)$$

where $\kappa = 8\pi G$(c=1), G is the gravitational constant, g is the determinant of the metric $g_{\mu\nu}$ and R is the Ricci scalar. $f(R)$ is a function of R and in the present study we write $f(R)$ as $f(R) = R + \lambda R^2$, where λ is a constant.

To be compatible with special relativity, gravity must be causal, ie., any change to a gravitating source must be communicated to distant observers with a speed no faster than the speed of light. This leads immediately to the idea that there must exist some notion of "gravitational radiation"[**2**]. The objects massive and relativistic enough to generate detectable Gravitational Waves(GWs) are astrophysical. There are no direct observations of GWs till today. Of the detectors designed to search for GWs, interferometers and resonators are the simplest ones. "Spherical antenna detectors" whose origin date back to the 1970s [**3**] are receiving lot of attention even now.

The theoretical frame work for the detection of GWs from GTR using spherical antennas had already been well developed. This paper deals with the production of GWs as a result of using the modified action (1). A massive mode in addition to the massless mode appears here. The massive mode of Gravitational Waves(GWs) could be thought of as a suitable candidate for dark matter present in galaxies. The feasibility of spherical antenna detection of massive GWs is also discussed.

2. Production of GWs

In order to find out the GWs arising from $f(R)$ theory, the modified action (1) is varied with respect to the metric alone and the resulting field equation is given by

$$f'(R)R_{\mu\nu} - \tfrac{1}{2}f(R)g_{\mu\nu} - (\nabla_\mu\nabla_\nu - g_{\mu\nu}\Box)f'(R) = \kappa T_{\mu\nu}. \qquad (2)$$

Substituting for $f(R)$ in (2) and taking trace we find,

$$6\lambda\Box R - R = \kappa T. \qquad (3)$$

Using the linearization condition $g_{\mu\nu} = \eta_{\mu\nu} + h_{\mu\nu}$ and also making use of trace of the reversed perturbation $\bar{h}_{\alpha\beta} = h_{\alpha\beta} - \frac{1}{2}\eta_{\alpha\beta}h$ and using the De Donder gauge $\partial^\beta \bar{h}_{\alpha\beta} = 0$, we get

$$-6\lambda\Box^2\bar{h}_{\mu\nu} + \Box\bar{h}_{\mu\nu} = \kappa T_{\mu\nu}. \qquad (4)$$

Considering the vacuum situation, we get two solutions

$$\bar{h}_{\mu\nu} = \int A(k)\exp(-i(k.r - \omega t))dk, \qquad (5)$$

$$\bar{h}_{\mu\nu} = \int A(k)\exp(-i(k.r - \omega_m t))dk. \qquad (6)$$

The general solution is a sum of solutions (5) and (6) and $\omega \neq \omega_m$. Equation (5) is the one that we usually get from general theory of relativity. Thus we have the usual general relativity solution and an additional solution containing the term m^2. Also in this case we have $k^\mu k_\mu = m^2$. Thus k^μ, the propagation vector, is no longer a null vector. Hence we cannot expect the massive GWs to travel with the speed of light.

3. Detection of massive modes using spherical antenna

Antennas generally consisted of an elastic body which may become deformed when gravitational radiation falls on it and thereby its normal modes get excited. Such an antenna measures precisely the Fourier transform of certain components of the Riemann curvature tensor averaged over its volume [**3**]. The present calculations of spherical antenna detection are done following Lobo [**4**]and Gasparini[**5**]. The five fundamental quadrupole modes of a spherical antenna are given by

$$\ddot{h}^\alpha(t) = \tfrac{8\pi}{15}M^\alpha_{ij}R_{i0j0}, \qquad (7)$$

with M^α_{ij} given by $M^{1c}_{ij} = \sqrt{\frac{15}{16\pi}}\begin{pmatrix} 0 & 0 & 1 \\ 0 & 0 & 0 \\ 1 & 0 & 0 \end{pmatrix}$, $M^{2c}_{ij} = \sqrt{\frac{15}{16\pi}}\begin{pmatrix} 1 & 0 & 0 \\ 0 & -1 & 0 \\ 0 & 0 & 0 \end{pmatrix}$,

$$M^{1s}_{ij} = \sqrt{\tfrac{15}{16\pi}}\begin{pmatrix} 0 & 0 & 0 \\ 0 & 0 & 1 \\ 0 & 1 & 0 \end{pmatrix}, M^{2s}_{ij} = \sqrt{\tfrac{15}{16\pi}}\begin{pmatrix} 0 & 1 & 0 \\ 1 & 0 & 0 \\ 0 & 0 & 0 \end{pmatrix}, M^{0}_{ij} = \sqrt{\tfrac{15}{16\pi}}\begin{pmatrix} -1 & 0 & 0 \\ 0 & -1 & 0 \\ 0 & 0 & 2 \end{pmatrix}.$$

By finding R_{i0j0}, $\ddot{h}^\alpha(t)$ for each mode α can be obtained. Then, the energy stored in the mode α at frequency ω_{nl} is given by [**5**]

$$E_s(n, l, \alpha) = \tfrac{1}{2}a_n^2\delta_{l,z}\left|\int \ddot{h}^\alpha(t)e^{-i\omega_{n2}t}dt\right|^2, \qquad (8)$$

where a_n is some quantity independent of t. n is a positive integer which represents the energy level for a fixed angular momentum l. It is found that the mode with $l = 2$ alone contributes. If we fix the value of n and replace $\ddot{h}^\alpha$, the energy stored for each mode α and hence the average energy stored can be found out. The rate of energy stored can be found out by calculating the ratio of energy stored in each mode to the average energy stored. This gives the energy sensitivity of the detector(antenna).

The metric perturbation from (5) and (6) can be written as[**6**]
$h_{\mu\nu}(t,z) = h_+(t-z)e^{(+)}_{\mu\nu} + h_\times(t-z)e^{(\times)}_{\mu\nu} + h_s(t-v_g z)e^s_{\mu\nu}$,
with $e^{(+)}_{\mu\nu} = \frac{1}{\sqrt{2}}\begin{pmatrix} 1 & 0 & 0 \\ 0 & -1 & 0 \\ 0 & 0 & 0 \end{pmatrix}$, $e^{(\times)}_{\mu\nu} = \frac{1}{\sqrt{2}}\begin{pmatrix} 0 & 1 & 0 \\ 1 & 0 & 0 \\ 0 & 0 & 0 \end{pmatrix}$, $e^s_{\mu\nu} = \frac{1}{\sqrt{2}}\begin{pmatrix} 0 & 0 & 0 \\ 0 & 0 & 0 \\ 0 & 0 & 1 \end{pmatrix}$.
Riemann tensor from our modified theory may be written as [**7**]

$$R_{i0j0} = \frac{1}{2}\begin{pmatrix} \ddot{h}_+ & \ddot{h}_\times & 0 \\ \ddot{h}_\times & -\ddot{h}_+ & 0 \\ 0 & 0 & m_g^2 h_s \end{pmatrix}_{ij}.$$

Converting the Riemann tensor to the detector frame, for a generic arrival direction given by the angle (θ, ϕ), the components of $\ddot{h}^\alpha(t)$ can be written as [**5**]

$$\ddot{h}^0(t) = \sqrt{\tfrac{\pi}{5}}(\sin^2\theta\ddot{h}_+ + m_g^2\cos^2\theta h_s),$$

$$\ddot{h}^{1c}(t) = \sqrt{\tfrac{4\pi}{15}}(-\sin\theta\cos\theta\cos\phi\ddot{h}_+ + \sin\theta\sin\phi\ddot{h}_\times - m_g^2\sin\theta\cos\theta\cos\phi h_s),$$

$$\ddot{h}^{1s}(t) = \sqrt{\tfrac{4\pi}{15}}(\sin\theta\cos\theta\sin\phi\ddot{h}_+ - \sin\theta\cos\phi\ddot{h}_\times + m_g^2\sin\theta\cos\theta\sin\phi h_s),$$

$$\ddot{h}^{2c}(t) = \sqrt{\tfrac{4\pi}{15}}(\cos^2\theta\cos^2\phi\ddot{h}_+ - 2\cos\theta\sin\phi\cos\phi\ddot{h}_\times - \sin^2\phi\ddot{h}_+ + m_g^2\sin^2\theta\cos^2\phi h_s),$$

$$\ddot{h}^{2s}(t) = \sqrt{\tfrac{4\pi}{15}}(\cos^2\theta\sin^2\phi\cos^2\phi\ddot{h}_+ - \cos\theta\sin^2\phi\ddot{h}_\times + \sin\phi\cos\phi\ddot{h}_+ + m_g^2\sin^2\theta\sin\phi\cos\phi h_s). \quad (9)$$

Then for a source emitting randomly polarized radiation, the average energy for each mode can be written as

$$\bar{E}_s(n,2,0) = \tfrac{\pi}{15}a_n^2\left[\tfrac{3}{4}\sin^4\theta\omega_{n2}^4(|\tilde{e}_+|^2 + |\tilde{e}_\times|^2) + m_g^4\cos^4\theta\,|\tilde{e}_s|^2\right],$$

$$\bar{E}_s(n,2,1c) = \tfrac{\pi}{15}a_n^2\left[\omega_{n2}^4(\sin^2\theta(\cos^2\theta\cos^2\phi + \sin^2\phi)(|\tilde{e}_+|^2 + |e_\times|^2) + (m_g^4\sin^2\theta\cos^2\theta\cos^2\phi\left|\tilde{h}_s(\omega_{n2})\right|^2)\right],$$

$$\bar{E}_s(n,2,1s) = \tfrac{\pi}{15}a_n^2[\omega_{n2}^4(\cos^2\theta\sin^2\phi+)(|\tilde{e}_+|^2 + |\tilde{e}_\times|^2) + (m_g^4\sin^2\theta\cos^2\theta\sin^2\phi\,|\tilde{e}_s|^2],$$

$$\bar{E}_s(n,2,2c) = \tfrac{\pi}{15}a_n^2[\omega_{n2}^4((1+\cos^2\theta)^2(\cos^2\phi - \tfrac{1}{2})^2 + 4\cos^2\theta\sin^2\phi\cos^2\phi))$$
$$(|\tilde{e}_+|^2 + |e_\times|^2) + (m_g^4\sin^4\theta\,|\tilde{e}_s|^2)],$$

$$\bar{E}_s(n,2,2s) = \tfrac{\pi}{15}a_n^2[\omega_{n2}^4((1+\cos^2\theta)^2\cos^2\phi\sin^2\phi + \cos^2\theta(\cos^2\phi - \sin^2\phi)^2)$$
$$(|\tilde{e}_+|^2 + |\tilde{e}_\times|^2) + (m_g^4\sin^4\theta\sin^2\phi\cos^2\phi\,|\tilde{e}_s|^2)], \quad (10)$$

with $h_+ = \cos(\psi)e_+ + \sin(\psi)e_\times$,
$h_\times = -\sin(\psi)e_+ + \cos(\psi)e_\times$,
$h_s = e_s$. ($h_s$has no ψ dependence). (11)

It is clear from these equations that the total average energy is not a constant [**8**]. It varies with θ and ϕ. Considering only those contributions from the massive mode, the total average energy is given by

$$\bar{E}_s(n,2,tot) = \tfrac{2\pi}{15}a_n^2 m_g^4[\tfrac{3}{4}\cos^2\theta + \cos^2\theta\sin^2\theta(\tfrac{1}{4} - \cos^2\phi) + \sin^2\theta\cos^2\phi]\left|\tilde{h}_s(\omega_{n2})\right|^2. \quad (12)$$

This indicates that the total sensitivity of the detector does not remain the same always and it depends on the GW direction. The variation of energy stored in each mode α with θ and ϕ are shown in the plots (Fig. 1). The energy sensitivity is defined as $\epsilon_\alpha = \frac{\bar{E}_s(n,2,\alpha)}{\bar{E}_s(n,2,tot)}$. The plots show a comparison of total energy transferred from GWs to the sphere between the + and × polarizations and the scalar polarization in the order of modes $1c$, $1s$, $2c$, $2s$, 0. The light portion(yellow) indicates a maximum and the dark portion(blue) indicates minimum. The case of + and × polarizations have already been done in the literature. Here, in the massive case, the maximum has been shifted. The region where it showed a maximum before is no more a region of maximum sensitivity. For instance, the rate ϵ_α for the mode $\alpha = 0$ has a maximum value $\frac{3}{4}$ in the massless case, whereas for the massive component, the maximum value is 1. It is then clear that the rate has been increased by 25% which can be recognized easily if GWs are detected using a spherical antenna. The total sensitivity for + and × polarizations was a constant no matter what the direction of the GWs is. But here the total sensitivity becomes direction dependent. We can see from the plots that the energy sensitivity of each mode has been increased compared to those of GTR.

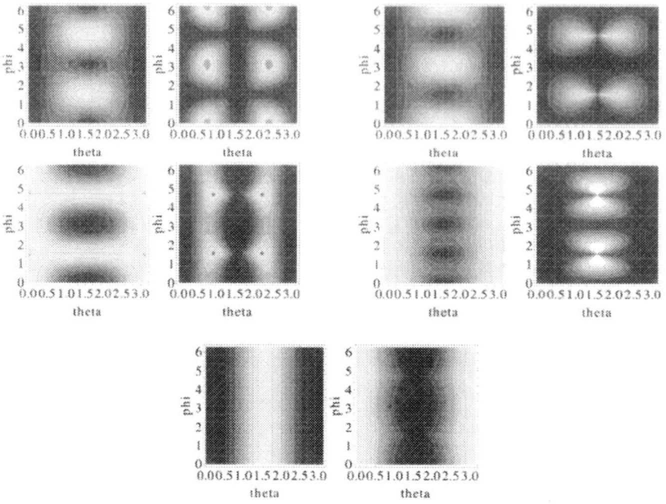

Figure 1. Comparision of ϵ_α of GW from GR(left) with that from f(R)(right) with ϵ_{1c}, ϵ_{1s}, ϵ_{2c}, ϵ_{2s}, ϵ_0 respectively from top left

4. Conclusion

In this paper, GWs from $f(R)$ gravity with $f(R)$ of the form $R + \lambda R^2$, λ being some unknown constant, is obtained. A massive mode in addition to the massless mode as in the case of GTR is obtained. This mode is found to propagate with a speed less than that of light. The detection of massive mode using spherical antenna detector has been analyzed. It showed an increase in energy sensitivity of the antenna. Also, the antenna is found to become highly directional towards such a massive component and if so, such a massive component can be easily recognizable and will be detectable if ever possible.

5. Acknowledgement

One of us (PP) would like to thank UGC, New Delhi for financial support through the award of a Junior Research Fellowship(JRF). VCK would like to acknowledge Associateship of IUCAA, Pune.

6. References

[1] Felice Antonio De and Tsujikawa S 2010 *Living Rev. Relativity* **13** 3
[2] Flanagan Eanna E and Hughes Scott A 2005 *New. J. Phys.* **7** 204
[3] Weber J 1969 *Phys. Rev. Lett.* **22** 1320
[4] Lobo J A 1995 *Phys. Rev* D **52** 591
[5] Gasparini Maria Alice 2005 *Phys. Rev* D **72** 104012
[6] Capozziello Salvatore and Faraoni Valerio 2011 *Beyond Einstein Gravity* (New York: Springer)
[7] Corda C 2008 *Astrophys. Space Sci.* **317** 95–106
[8] Rakhmanov Malik 2005 *Response of LIGO to gravitational waves at high freuencies and in the vicinity of the FSR(37.5kHz)* 060237–00

Journal of Physics: Conference Series **405** (2012) 012020 doi:10.1088/1742-6596/405/1/012020

Emergent gravity: From statistical point of view

Bibhas Ranjan Majhi
IUCAA, Post Bag 4, Ganeshkhind, Pune University Campus, Pune 411 007, India
E-mail: bibhas@iucaa.ernet.in

Abstract. Near the event horizon of a black hole, the effective theory is two dimensional conformal theory. Here we show that the holographic modes characterising this underlying conformal symmetry and the basic definition of entropy S in statistical mechanics lead the equipartition law of energy. We also show that S is proportional to the gravity action which suggests the emergent nature of gravity. This is further bolstered by expressing the generalised Smarr formula as a thermodynamic relation, $S = E/2T$, where T is the Hawking temperature and E is shown to be the Komar energy.

1. Introduction

In absence of a true quantum theory of gravity, semi-classical approaches have become increasingly popular and are widely used, particularly in the context of thermodynamics of gravity. It is now evident that gravity and thermodynamics are closely connected to each other [1, 2, 3]. The repeated failure to quantise gravity led to a parallel development [4, 5, 6] where gravity is believed to be an emergent phenomenon just like thermodynamics and hydrodynamics instead treating it as a fundamental force. In this context, the fundamental role of gravity is replaced by thermodynamical interpretations leading to similar or equivalent results without knowing the underlying microscopic details. Nevertheless, understanding the entropic or thermodynamic origin of gravity is far from complete since the arguments are more heuristic than concrete and depend upon specific ansatz or assumptions.

In this writeup, based on our work [7], we readdressed some of the facts related to the emergent nature of gravity, in the context of the black holes in Einstein gravity.

Here using certain basic results derived by us [8, 9] and applying the standard definition of entropy given in statistical mechanics, we are able to provide a statistical interpretation of gravity. We first give the equipartition law of energy and show that this leads to the identification of entropy with the action for gravity. The immediate consequence of it is that the Einstein equations, obtained by a variational principle involving the action, can be equivalently obtained by an extremisation of the entropy. This implies that the gravity can be thought of as the emergent phenomenon.

The emergent nature has been further bolstered by deriving the relation $S = E/2T$, connecting the entropy (S) with the Hawking temperature (T) and energy (E) for a black hole with stationary metric. We show that this energy corresponds to Komar's expression [10, 11]. Using this fact we show that the relation $S = E/2T$ is also compatible with the generalised Smarr formula [12, 3, 13].

2. Previous results

Here we briefly introduce three relevant results, which were derived previously by some of us, for our main purpose.

• Thermodynamics of a black hole is universally governed by its properties near the event horizon. It is also well understood that near the event horizon the effective theory becomes two dimensional conformal theory whose metric is given by the two dimensional $(t - r)$- sector of the original metric. For details see [14].

• Using the WKB approximation, the left (L) and right (R) moving (holomorphic) modes are obtained by solving the appropriate field equation under the effective two dimensional metric. It has been shown that the modes inside and outside the horizon are related by the transformations [8, 9]:

$$\phi_{in}^{(R)} = e^{-\frac{\pi\omega}{\kappa}}\phi_{out}^{(R)}; \quad \phi_{in}^{(L)} = \phi_{out}^{(L)} \tag{1}$$

where "ω" is the energy of the particle as measured by an asymptotic observer and "κ" is the surface gravity of the black hole. Furthermore, the L mode gets trapped while the R mode tunnels through the horizon and is observed at asymptotic infinity as Hawking radiation [8, 9]. The probability of this "R" mode, to go outside, as measured by the outside observer is given by

$$P^{(R)} = \left|\phi_{in}^{(R)}\right|^2 = \left|e^{-\frac{\pi\omega}{\kappa}}\phi_{out}^{(R)}\right|^2 = e^{-\frac{2\pi\omega}{\kappa}} \tag{2}$$

where, in the second equality, (1) has been used. This is essential since the measurement is done from outside and hence $\phi_{in}^{(R)}$ has to be expressed in favour of $\phi_{out}^{(R)}$.

• The effective two dimensional curved metric can be embedded in a flat space which has exactly two space-like coordinates. This is a consequence of a modification in the original GEMS (globally embedding in Minkowskian space) approach of [15] and has been elaborated by us in [16]. This tells that each R mode can be associated to two degrees of freedom. Therefore, the total number of degrees of freedom for n number of R modes is $N = 2n$.

3. Partition function

The partition function for the space-time with matter field is given by [13],

$$\mathcal{Z} = \int D[g, \Phi]\ e^{iI[g,\Phi]} \tag{3}$$

where $I[g, \Phi]$ is the action representing the whole system and $D[g, \Phi]$ is the measure of all field configurations (g, Φ). Since we want to confine ourself within the usual semi-classical regime, we shall neglect all the higher order terms for the subsequent analysis. Therefore, in the semi-classical regime the partition function is expressed as [13],

$$\mathcal{Z} \simeq e^{iI[g_0,\Phi_0]}, \tag{4}$$

where (g_0, Φ_0) are the background fields and the classical action $I[g_0, \Phi_0]$ leads to the Einstein equation.

4. Relation between different thermodynamical quantities with the action and their implications

Here using the previous results and the above definition of partition function combined with the standard definition of entropy in statistical mechanics we derive three important relations among

Journal of Physics: Conference Series **405** (2012) 012020 doi:10.1088/1742-6596/405/1/012020

different thermodynamical quantities of a stationary black hole. The physical implications are also being described at the end.

In statistical mechanics the entropy is related to the partition function by the relation: $S = \ln \mathcal{Z} + \frac{E}{T}$ and so (4) leads to, $S = iI[g_0, \Phi_0] + \frac{E}{T}$ where E and T are respectively the energy and temperature of the system. Now since there are N number of degrees of freedom in which all the information is encoded, the entropy (S) of the system must be proportional to N. Hence

$$N = N_0 S = N_0 (iI[g_0, \Phi_0] + \frac{E}{T}), \tag{5}$$

where N_0 is a proportionality constant. Now the average value of the energy, measured from outside, is calculated as,

$$< \omega >= \frac{\int_0^\infty d\omega\ \omega\ P^{(R)}}{\int_0^\infty d\omega\ P^{(R)}} = T \tag{6}$$

where $T = \kappa/2\pi$ is the temperature of the black hole [9]. Therefore if we consider that the energy E of the system is encoded near the horizon and the total number of pairs created is n among which this energy is distributed, then we must have, $E = nT$ where only the R mode of the pair is significant. Hence, since $N = 2n$, we obtain the energy of the system as

$$E = \frac{1}{2} NT. \tag{7}$$

Noted that (7) can be interpreted as the usual *law of equipartition of energy*, since it implies that if the energy E is distributed equally over each degree of freedom, then each degree of freedom should contain an energy equal to $T/2$. The fact that the energy is equally distributed among the degrees of freedom may be understood from the symmetry of two space-like coordinates ($z^1 \longleftrightarrow z^2$) such that the metric is unchanged [16].

Now using (5) and (7) we obtain following relations:

$$E = \frac{N_0}{2 - N_0} iTI[g_0, \Phi_0]; \quad S = \frac{2E}{N_0 T}. \tag{8}$$

In order to fix the value of "N_0" we consider the simplest example, the Schwarzschild black hole for which the entropy, energy and temperature are given by, $S = \frac{A}{4} = 4\pi M^2$, $E = M$, $T = \frac{1}{8\pi M}$, where "M" is the mass of the black hole. Substitution of these in the second relation of (8) leads to $N_0 = 4$. Finally, putting back $N_0 = 4$ in the preceding relations we obtain,

$$E = -\frac{i\kappa I[g_0, \Phi_0]}{\pi}; \quad S = -iI[g_0, \Phi_0] = \frac{E}{2T}. \tag{9}$$

The implications of the above relations are as follows:

• Use of the explicit form of Einstein-Hilbert action in the first relation shows that E is the Komar conserved quantity corresponding to Killing vector. For details, see [7].

• First equality of the second relation signifies that the extremisation of entropy leads to Einstein's equations. It illustrates the emergent nature of gravity.

• The last equality $S = E/2T$ is the general expression for Smarr formula. This can be checked by substituting the relevant quantities of Kerr-Newman black hole to obtain the relation $\frac{M}{2} = \frac{\kappa A}{8\pi} + \frac{VQ}{2} + \Omega J$, which is the generalised Smarr formula [12, 3, 13]. This further clarifies the fact that the gravity is an emergent phenomenon.

Journal of Physics: Conference Series **405** (2012) 012020 doi:10.1088/1742-6596/405/1/012020

5. Conclusions

We, in the context of black hole, using the statistical definition of entropy and the partition function for gravity, further clarified the possibility of considering gravity as an emergent phenomenon. There are certain issues still to be addressed or clarified. Let us tabulate some of them:

• First of all, we saw that the proportionality constant N_0 was not fixed from the fundamental ground. Rather, we did it by using the parameters of a specific black hole. An independent derivation will be very much interesting.

• Is it possible to discuss the same for more general situation, like the spacetime without black hole? This may be done by considering the local Rindler frame in the spacetime which has a null surface on which the temperature and entropy can be associated.

• A more simpler but interesting extension would be inclusion of higher dimensions. In this regard, one can be interested to see if the higher dimensional gravity exhibit same feature or in other words if the above features depend on dimension of spacetime. Some part of this issue has already been addressed in [17]. Here we did for the higher dimensional black hole in Einstein gravity and showed that we again have a relation similar to $E = 2ST$, but E is interpreted as the Komar conserved quantity rather than the energy itself.

• As a final remark we feel that although our results were derived for Einstein gravity, the methods are general enough to include other possibilities like $f(R)$ gravity, Lancozs-Lovelock gravity etc.

References

[1] J. D. Bekenstein, Phys. Rev. D **7**, 2333 (1973).
[2] S. W. Hawking, Nature **248**, 30 (1974). S. W. Hawking, Commun. Math. Phys. **43**, 199 (1975) [Erratum-ibid. **46**, 206 (1976)].
[3] J. M. Bardeen, B. Carter and S. W. Hawking, Commun. Math. Phys. **31**, 161 (1973).
[4] T. Jacobson, Phys. Rev. Lett. **75**, 1260 (1995) [arXiv:gr-qc/9504004].
[5] T. Padmanabhan, Rept. Prog. Phys. **73**, 046901 (2010) [arXiv:0911.5004 [gr-qc]] and references therein.
[6] E. P. Verlinde, JHEP **1104**, 029 (2011) [arXiv:1001.0785 [hep-th]].
[7] R. Banerjee and B. R. Majhi, Phys. Rev. D **81**, 124006 (2010) [arXiv:1003.2312 [gr-qc]].
[8] R. Banerjee and B. R. Majhi, Phys. Rev. D **79**, 064024 (2009) [arXiv:0812.0497 [hep-th]]. R. Banerjee and B. R. Majhi, Phys. Lett. B **675**, 243 (2009) [arXiv:0903.0250 [hep-th]].
[9] R. Banerjee, B. R. Majhi and E. C. Vagenas, Phys. Lett. B **686**, 279 (2010) [arXiv:0907.4271 [hep-th]].
[10] A. Komar, Phys. Rev. **113**, 934 (1959).
[11] R. M. Wald, "General Relativity," *Chicago, Usa: Univ. Pr. (1984) 491p.*
[12] L. Smarr, Phys. Rev. Lett. **30**, 71 (1973) [Erratum-ibid. **30**, 521 (1973)].
[13] G. W. Gibbons and S. W. Hawking, Phys. Rev. D **15**, 2752 (1977).
[14] B. R. Majhi, arXiv:1110.6008 [gr-qc].
[15] S. Deser and O. Levin, Phys. Rev. D **59**, 064004 (1999) [arXiv:hep-th/9809159].
[16] R. Banerjee and B. R. Majhi, Phys. Lett. B **690**, 83 (2010) [arXiv:1002.0985 [gr-qc]].
[17] R. Banerjee, B. R. Majhi, S. K. Modak and S. Samanta, Phys. Rev. D **82**, 124002 (2010) [arXiv:1007.5204 [gr-qc]].

Journal of Physics: Conference Series **405** (2012) 012021 doi:10.1088/1742-6596/405/1/012021

Simulating matched filter detection of ionized bubble around a quasar in the epoch of reionization

Suman Majumdar[1], **Somnath Bharadwaj**[1] **and Tirthankar Roy Choudhury**[2]

[1] Department of Physics and Meteorology and Centre for Theoretical Studies, IIT Kharagpur, Kharagpur - 721302, India

[2] Harish-Chandra Research Institute, Allahabad - 211019, India

E-mail: sumanm@phy.iitkgp.ernet.in

Abstract. The recent discovery of the z = 7.085 quasar has opened up a new window to peep into the inter galactic medium during the epoch of reionization (EoR). The detection of the ionized region around such a high redshift quasar is in principle capable of constraining the neutral fraction of the IGM and the quasar's age. Here we present the possibility of detection of such an ionized region around a quasar at z = 8 using redshifted 21-cm observations of the neutral hydrogen and possible ways to constrain IGM neutral fraction and quasar's age through this detection.

1. Introduction

The epoch of reionization is one of the least known periods in the history of our Universe. According to the present understanding, reionization of neutral hydrogen (H I) is an extended process spanning over the redshift range $6 \lesssim z \lesssim 15$. The intergalactic medium (IGM) during this period is characterised by bubbles of ionized hydrogen (H II), centred around luminous sources (galaxies and quasars). Though very rare but the high redshift quasars can ionize very large volumes of the IGM around them with respect to their stellar counterpart. Detection of these H II bubbles will directly probe the state of the local IGM around the ionizing sources. It will also constrain the properties of the quasar, such as it's luminosity and age.

The highest redshift at which a quasar has been detected till date is $z = 7.085$ (Mortlock et al., 2011). The Ly-α absorption spectrum of this object reveals a highly ionized near zone of radius 2.1 ± 0.1 Mpc (physical). (Bolton et al., 2011) have constrained the τ_Q and $\langle x_{H\ I} \rangle_V$ using simulated Ly-α absorption spectra of this quasar. This study shows that there are several combinations of the τ_Q and $\langle x_{H\ I} \rangle_V$ all of which can reproduce the observed data. The constrains thus obtained over $\langle x_{H\ I} \rangle_V$ and τ_Q are very wide. Ly-α absorption spectra are pencil beam observation along a line of sight (LoS) and they get saturated for a $x_{H\ I} > 10^{-4}$, because of these reasons they are limited in their ability to determine the age of the quasar and the local neutral fraction.

The redshifted 21-cm emission from neutral hydrogen in the epoch of reionization is believed to be a powerful tool to detect H II bubbles around quasars. The 21-cm intensity is directly proportional to the H I density, and it is in principle possible to probe the entire ionization profile of the H II bubble. This is particularly motivated by the fact that the presently functioning

GMRT, LOFAR and upcoming MWA[1] are all sensitive to the H I signal from the EoR. However, redshifted H I signal is extremely small relative to the sensitivities of the present and upcoming telescopes and it will be buried deep in foregrounds which are a few orders of magnitude larger than the signal.

Datta, Bharadwaj & Choudhury (2007) (hitherto Paper I) have proposed a matched filter technique to detect ionized bubbles in radio-interferometric observations of redshifted 21-cm emission. The matched filter optimally combines the entire signal of an ionized bubble while minimizing the noise and the foreground contributions. This technique assumes the H II bubble to be spherical. However, a growing spherical bubble will appear anisotropic along the LoS to a present day observer due to the finite light travel time (FLTT) (Shapiro & Giroux, 1987; Wyithe & Loeb, 2004; Yu, 2005; Wyithe, Loeb & Barnes, 2005; Sethi & Haiman, 2008) and also due to the evolution of global ionization fraction (Geil et al., 2008). In an earlier work (Majumdar et al. (2011); hitherto Paper II), we have analytically quantified this anisotropy and studied the possibility of detecting such a bubble in a targeted search around a known quasar.

In this work we investigate the possibility of using such an observation to constrain the properties of the quasar and the IGM. For this we improve upon the spherical filter by calculating the apparent, anisotropic shape of the quasar bubble and using this as a template for the filter. We expect this to give a better match to the bubble that is actually present in the data. We also use a semi-numerical technique to simulate the quasar bubble while incorporating the effect of other H II bubbles due to stellar sources and density fluctuations in the IGM.

2. Simulation

For a isotropic and constant photon emission rate ($\dot{N}_{phs}$) of a quasar the solution to the growth model for the H II bubble around it takes the form

$$r(\tau) = r_S \left[1 - \exp\left(-\frac{\tau}{\tau_{rec}}\right)\right]^{\frac{1}{3}} \tag{1}$$

[eq. (2) of Paper II] where $\tau_{rec} = x_{H\ I}\left(C\langle n_H\rangle \alpha_B\right)^{-1}$ and $r_S = \left(3\dot{N}_{phs}\tau_{rec}/(4\pi x_{H\ I}\langle n_H\rangle)\right)^{1/3}$. The age of the quasar τ_Q is further related to the angle with the LoS ϕ by

$$\tau_Q = \tau + \frac{r(\tau)}{c}(1 - \cos\,\phi) \tag{2}$$

[eq. (7) of Paper II]. The simultaneous solution of these two equations provide the observed apparent shape of the bubble. The apparent anisotropy in shape is quantified by a parameter η (eq. (9) of Paper II). A value $\eta > 0$ indicates that the bubble is elongated along the LoS and a value $\eta < 0$, indicates that it is compressed along the LoS.

We have implemented the semi-numerical formalism proposed by Choudhury, Haehnelt & Regan (2009) for generating the ionization field at a given redshift. We have generated the dark matter distribution at $z = 8$ using a Particle Mesh Nbody code. The volume of the simulation is constrained by the 16 Gigabytes of memory available in our computer. We perform our simulation in a periodic box of size 85.12 Mpc (comoving) with 1216^3 grid points and 608^3 particles, with a mass resolution $M_{part} = 7.275 \times 10^7\,h^{-1}\,M_\odot$.

We identify halos within the simulation box using a standard Friend-of-Friend algorithm, with a fixed linking length $b = 0.2$ (in units of mean inter particle distance) and minimum halo mass $= 10M_{part}$. We assume that the ionizing luminosity from a galaxy is proportional to the mass of it's halo. In the semi-numerical formalism, a region is said to be ionized if the average

[1] http://www.gmrt.ncra.tifr.res.in; http://www.lofar.org/ and http://www.haystack.mit.edu/ast/arrays/mwa/

Journal of Physics: Conference Series **405** (2012) 012021 doi:10.1088/1742-6596/405/1/012021

number of photons reaching there exceeds average neutral hydrogen density at that point. We have used a 256^3 grid with a grid spacing of 0.3325 Mpc for simulating the ionization maps. The quasar is assumed to be located at the most massive halo and snapshots of H I maps at different stages of growth of it's bubble is generated. We then choose sections of the simulation box at different stages of the bubble's growth and stack them together to generate the apparent shape of the H II bubble as seen by a present day observer. We consider both uniform and non-uniform density dependent recombinations in our simulations. We choose three distinct stages of anisotropy to generate the quasar bubble's apparent shape and $\dot{N}_{phs}$ and τ_Q are adjusted so that the bubble can be accommodated within the simulation box. We choose the background mass averaged neutral fraction of the IGM to be $\langle x_{H\ I} \rangle_M = 0.5$ at $z = 8$ and to remain constant during the growth of the bubble. Figure 1 shows the simulated quasar bubble at very early ($\eta = 0.05$), middle ($\eta = -0.13$) and very late ($\eta = -0.02$) stages of it's growth.

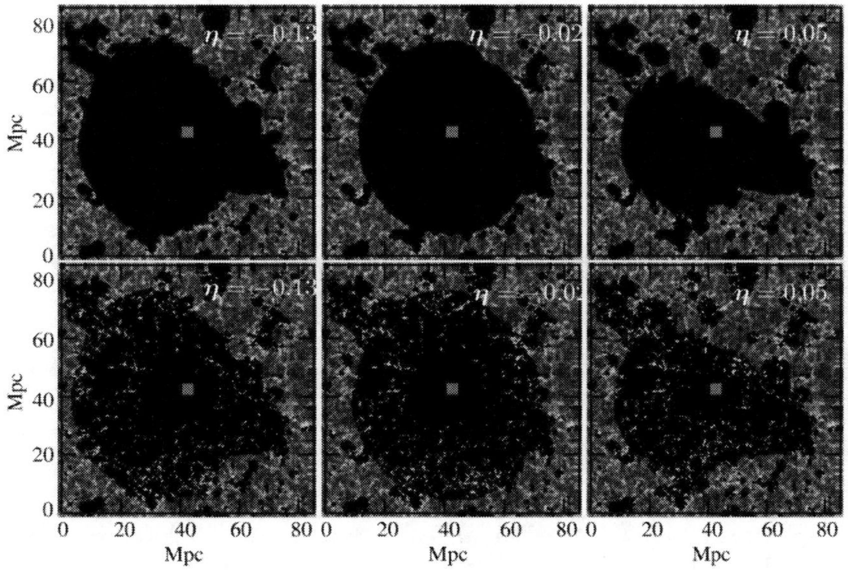

Figure 1. This shows the simulated quasar bubble at different stages of it's growth in one realization of our simulations. Top and bottom panels show the shapes considering uniform and non-uniform recombinations respectively. Red square shows the location of the quasar in each panel.

3. Bubble detection and parameter estimation

HI maps are then converted into visibilities to calculate the estimator for matched filter search as in Paper I.

$$\hat{E} = \sum_{a,b} S_f^*(\vec{U}_a, \nu_b)\hat{\mathcal{V}}(\vec{U}_a, \nu_b) \tag{3}$$

Here the filter $S_f(\vec{U}, \nu)$ (see Paper I for details) is the signal expected from the bubble that we are searching for and $\hat{\mathcal{V}}(\vec{U}, \nu)$ are the measured visibilities. All the contributions to $\mathcal{V}$, except the signal $S(\vec{U}, \nu)$, are assumed to be random variables of zero mean, uncorrelated to the filter. In a targeted search the sky plane position of the quasar will be known. The parameters which control the bubble as well as the anisotropic filter's shape are $\dot{N}_{phs}/C$, $x_{H\ I}/C$ and τ_Q. Among them, $\dot{N}_{phs}$ can be estimated from the Ly-α absorption spectra of the quasar and the clumping factor C from simulations. In the matched filter technique we expect the signal to noise ratio (SNR)

$$\text{SNR} = \frac{\langle \hat{E} \rangle}{\sqrt{\langle (\Delta \hat{E})^2 \rangle}} \tag{4}$$

to peak when the filter parameters ($x_{H\ I}$ and τ_Q) exactly match with that of the bubble. We use eq. (19) of Paper I to calculate the noise contribution to the variance of the estimator. We propose that the best matched filter parameters will give an estimate of $x_{H\ I}$ and τ_Q. Figure 2 shows the estimates of $x_{H\ I}$ and τ_Q obtained from six different realizations of our

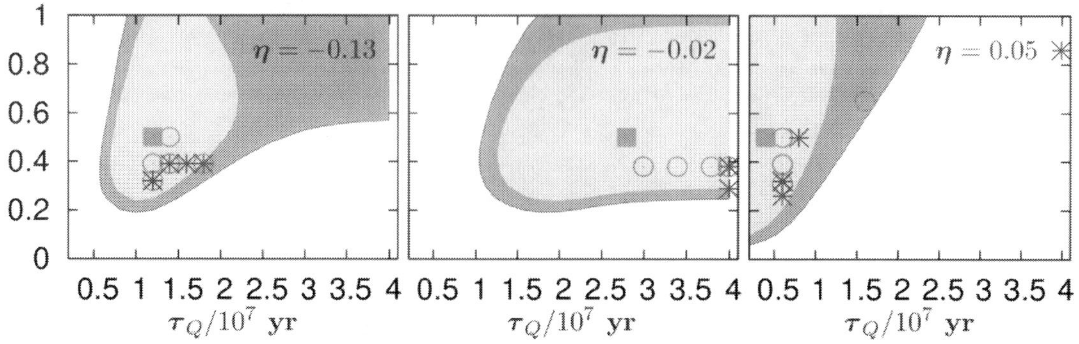

Figure 2. This shows the estimated quasar and IGM parameters from six different realizations of our simulations. Circles and stars represent search results from uniform and non-uniform recombination scenarios respectively. Pink square shows actual values of input parameters. Blue and green shaded regions show the uncertainties ($\Delta SNR = 1$) in the estimated parameters for 4000 and 9000 hrs of observations with GMRT.

simulations considering both uniform and non-uniform recombinations in the IGM. We observe that in most of the realizations the estimates are well within the uncertainties predicted for 9000 hrs of observations with GMRT. At the early stage of growth ($\eta = +$ve) when the bubble appears elongated along the LoS, one can only obtain a lower limit on $x_{H\ I}$ even for 9000 hrs of observation but precise lower and upper bounds on τ_Q can be marked. On the contrary at very late stages of growth ($\eta \sim 0$) one can precisely measure $x_{H\ I}$ but can only obtain a lower limit on τ_Q. The middle stage of growth ($\eta =$ -ve) when the bubble appears compressed along LoS appears to be the most suitable for parameter estimation, and lower and upper bounds on both $x_{H\ I}$ and τ_Q can be obtained at this stage. We also observe that in almost all realizations the $x_{H\ I}$ is slightly underestimated from its actual value. Clustering of stellar sources around the quasar is the main reason for this underestimation.

References

Bolton, J. S., Haehnelt, M. G., Warren, S. J., Heweet, P. C., Mortlock, D. J., Venemans, B. P., McMahon, R. G., Simpson, C. 2011, *MNRAS*, **416**, L70

Choudhury, T. R., Haehnelt, M. G., & Regan, J. 2009, *MNRAS*, **394**, 960

Datta, K. K., Bharadwaj, S., & Choudhury, T. R., 2007,*MNRAS*, **382**, 109

Geil, P. M., Wyithe, J. S. B., Petrovic, N., & Oh, S. P. 2008, *MNRAS*, **390**, 1496

Majumdar, S., Bharadwaj, S., Datta, K., K., & Choudhury, T. R. 2011, *MNRAS*, **413**, 1409

Mortlock, D., J., et al. 2011, *Nature*, **474**, 7353

Sethi, S., & Haiman, Z. 2008, *AJ*, **673**, 1S

Shapiro, P. R., & Giroux, M. L. 1987, *ApJL*, **321L**, 107S

Wyithe, J. S. B., & Loeb, A. 2004, *ApJ*, **610**, 117

Wyithe, J. S. B., Loeb, A., & Barnes, D. G. 2005, *ApJ*, **634**, 715

Yu, Q. 2005, *ApJ*, **623**, 683

Journal of Physics: Conference Series **405** (2012) 012022 doi:10.1088/1742-6596/405/1/012022

Preheating Mechanism in F-term SUSY-Hybrid Inflation

Arindam Mazumdar

Theory Division, Saha Institute of Nuclear Physics, 1/AF-Bidhannagar, Kolkata-64, India

E-mail: arindam.mazumdar@saha.ac.in

Abstract. Supersymmetric F-term hybrid inflation is one of the most popular models of inflation. Preheating process occurs in this model via two different mechanism. Firstly the standard parametric resonance and secondly, the tachyonic preheating. Generally tachyonic preheating dominates the parametric resonance for this type of models. For different values of the parameters of the theory dominance of tachyonic preheating can vary.

1. Introduction

Inflationary paradigm has been greatly successful in solving the horizon problem and flatness problem and describing the origin of observed cosmic microwave background (CMB) anisotropy. While shorting out the best model for inflation is still a very active area of modern cosmology, theory for describing matter formation from the energy of inflation is still not well developed. Since an exponential expansion at the time of inflation extremely decreases the temperature of the universe a process of reheating is necessary to get present observed CMB temperature. But a standard reheating process dominated by the elementary decay of inflaton field to the other matter fields is not sufficient enough to produce proper reheating temperature. In 90's a nonlinear and non-perturbative theory of rapid production of matter field from the inflaton field got developed, which helped to reduce the time of long matter dominated era just after the end of inflation[1]. This rapid growth of matter fields expected to happen prior to the elementary reheating, that is why this process is known as *preheating*.

In this particular note we will be concentrating on the F-term SUSY Hybrid model of inflation. From WMAP data it became prominent that scalar spectral index is less than one. So standard hybrid inflation was replaced by its supersymmetric version to match with the observation. Specially F-term SUSY Hybrid model is tailor-made to fit with Higgs mechanism.

Preheating mechanism in this model is of two types, tachyonic preheating and parametric resonance. In general tachyonic preheating is more fast and effective than parametric resonance. But depending on the parameters of the model and comoving wavenumber of the matter field one process may be favored over other.

2. F-term SUSY Hybrid Model

In general F-term SUSY-Hybrid inflation assumes a superpotential of the following form[2].

$$W = \kappa\Phi(\Psi\bar{\Psi} - M^2) \tag{1}$$

Journal of Physics: Conference Series **405** (2012) 012022 doi:10.1088/1742-6596/405/1/012022

Here $\Psi, \bar{\Psi}$ are chiral scalar multiplets conjugate to each other. Φ is a gauge singlet scalar and contains inflaton (ϕ). F-term scalar potential coming from this super-potential is

$$V_F = \kappa^2(|\psi|^2 - M^2)^2 + 2\kappa^2|\phi|^2|\psi|^2. \tag{2}$$

where ψ and $\bar{\psi}$ can be any scalar multiplet conjugate to each other. Inflation occurs in flat direction $\psi = \bar{\psi} = 0$ which has a constant potential of $\kappa^2 M^4$. Total effective potential by driving the inflation is the sum of this constant part and one loop correction in this direction[3, 4]:

$$\begin{aligned} V(\phi) &= \kappa^2 M^4 + \frac{\kappa^4 \mathcal{N}}{32\pi^2}\left((\phi^2 + M^2)^2 \log\frac{\kappa^2(\phi^2 + M^2)}{\Lambda^2}\right. \\ &\left. + (\phi^2 - M^2)^2 \log\frac{\kappa^2(\phi^2 - M^2)}{\Lambda^2} - 2\phi^4 \log\frac{\kappa^2\phi^2}{\Lambda^2}\right) \end{aligned} \tag{3}$$

where $\mathcal{N}$ is the dimensionality of the representation of $\psi, \bar{\psi}$ and Λ is the renormalization scale. This potential has two minina, one is the local minima at $\phi = 0, \psi = 0$ and another is the global minima at $\phi = 0, \psi = M$. After the end of inflation another susy-breaking term, say $m^2\phi^2$ is required to drive inflaton to $\phi = 0$. Value of m^2 is taken much smaller than $\kappa^4 M^2$.

So, generally these types of potentials can be written as

$$V(\phi) = V_0 + v(\phi). \tag{4}$$

where V_0 is the constant vacuum energy density and $v(\phi)$ is the loop correction term.

Scalar spectral index(n_s) and amplitude of density perturbation ($\Delta_{\mathcal{R}}$) of CMB are related to the inflation potential through the slowroll parameters(ϵ, η).

$$n_s = 1 - 6\epsilon + 2\eta \tag{5}$$

$$\Delta^2_{\mathcal{R}} = \frac{8}{m^4_{Pl}} \frac{V}{\epsilon}\bigg|_{k=aH} \tag{6}$$

where,

$$\epsilon = \frac{m^2_{Pl}}{16\pi}\left(\frac{V_\phi}{V}\right)^2 = \frac{m^2_{Pl}}{16\pi}\left(\frac{v_\phi}{V_0 + v}\right)^2, \tag{7}$$

$$\eta = \frac{m^2_{Pl}}{8\pi}\left(\frac{V_{\phi\phi}}{V}\right) = \frac{m^2_{Pl}}{8\pi}\left(\frac{v_{\phi\phi}}{V_0 + v}\right). \tag{8}$$

Here ϕ in subscript means derivative with respect to ϕ and $m_{Pl} = 1/\sqrt{G}$.

Amplitude of tensor perturbation is related to inflation potential as

$$\Delta^2_h = \frac{128}{3m^4_{Pl}} V(\phi)\bigg|_{k=aH} \tag{9}$$

So tensor-to-scalar ratio takes the form $r = \frac{\Delta^2_h}{\Delta^2_{\mathcal{R}}} = 16\epsilon \approx \frac{128}{3m^4_{Pl}\Delta^2_{\mathcal{R}}}\kappa^2 M^4$. SUSY hybrid model is a model with negative η, which gives negligibly small tensor-to-scalar ratio. If we take number of *e-foldings*, N, corresponding to the WMAP pivot scale 0.002 MPc^{-1}, to be 60 and $\mathcal{N}$ is 8 then for $\Delta^2_{\mathcal{R}} = (2.41 \pm 0.11) \times 10^{-9}$ and $n_s = 0.983$ [5]we should have $M = 9.55 \times 10^{15}$GeV and $\kappa = 0.008 - 0.01$.

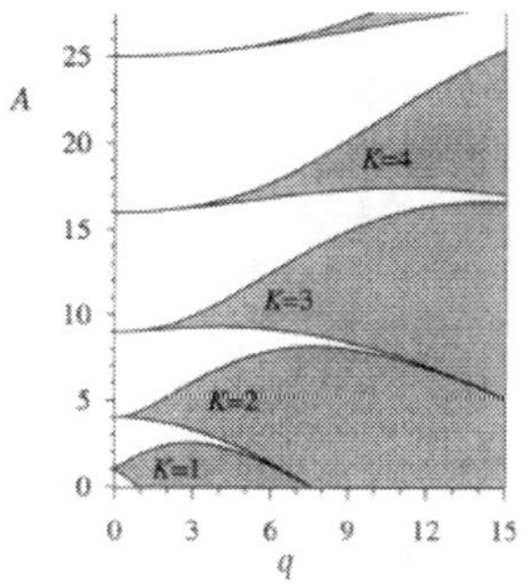

Figure 1. Mathieu chart showing stable(white) and unstable(dark) region in A, q space.[6]

3. Parametric Resonance Mechanism

We Fourier decompose the spatial part of ψ and derive the equations of motion from equation(2). After taking mean-field approximation we get

$$\ddot{\psi}_k + 3H\dot{\psi}_k + \frac{k^2}{a^2}\psi_k + 4\kappa^2\langle\psi^2\rangle\psi_k + 4\kappa^2\phi^2\psi_k - 4\kappa^2M^2\psi_k = 0 \tag{10}$$

$$\ddot{\phi} + 3H\dot{\phi} + 2\kappa^2\langle\psi^2\rangle\phi + m^2\phi = 0 \tag{11}$$

As usual inflaton ϕ is regarded as spatially homogeneous. $\langle\psi^2\rangle$ is the spatially averaged value of ψ^2 and k is the comoving wavenumber. By taking $X_k = a^{3/2}\psi_k$ we re-write equation(10) as:

$$\ddot{X}_k + \left[\frac{k^2}{a^2} + 4\kappa^2\phi^2 + 4\kappa^2\frac{\langle X^2\rangle}{a^3} - 4\kappa^2M^2\right]X_k = 0 \tag{12}$$

Here pressure term has been neglected since at the time of free oscillation it is vanishingly small. For oscillatory case of ϕ we assume $\phi = \phi_0 \cos\omega_\phi t$ and change the variable t to $\tau = \omega_\phi t$. In this way we arrive at the familiar form of Mathieu equation[7] from (10)

$$\frac{d^2X_k}{d\tau^2} + [A_k + 2q\cos(2\tau)]X_k = 0 \tag{13}$$

where

$$A_k = \left(k^2/a^2 + 4\kappa^2\frac{\langle X^2\rangle}{a^3} - 4\kappa^2M^2\right)/\omega_\phi^2 + 2q \tag{14}$$

$$q = \kappa^2\phi_0^2/\omega_\phi^2 \tag{15}$$

Solution of Mathieu equation takes the form like $X_k \sim e^{\nu_k\tau}$ where ν_k is the critical exponent and depends on A_k and q. In an unstable band of Mathieu chart (fig. 1) ν gets a real value which gives exponential growth in X_k. Occupation number of ψ_k is proportional to $|X_k^2| \sim e^{2\nu_k\tau}$[8]. From numerical solution of eqn.(10) occupation number can also be estimated using the the well known formula[7]

$$n_k = \frac{\omega_k}{2}\left(\frac{|\dot{\psi}_k|^2}{\omega_k^2} + |\psi_k|^2\right) - \frac{1}{2} \tag{16}$$

where $\omega_k^2 = \frac{k^2}{a^2} - 4\kappa^2M^2$.

In an expanding universe parametric resonance is always expected to be in broad resonance regime. A_k and q both change with time and parametric resonance is not constrained in a single unstable band of Mathieu chart (fig. 1). By observing the growth of n_k we can find the average critical exponent[9].

4. Tachyoninc Preheating Mechanism

Tachyonic preheating is a process in which if a field has tachyonic instability around some extrema, its amplitude can increase explosively and preheating gets complete within a very short time. In our model as the inflaton ϕ crosses the critical value M, potential for ψ becomes tachyonic. Amplitude of ψ_k keeps increasing as long as $k^2 + 4\kappa^2\psi_k^2 + 4\kappa^2\phi^2 - 4\kappa^2M^2$ remains negative. Occupation number of the ψ_k can be defined as equation(16) using $\omega_k^2 = \frac{k^2}{a^2}$ without any loss of generality [10]. If initially $\langle\psi^2\rangle$ is small compared ϕ and M, we can have

$$\psi_k \sim \exp(t\sqrt{4\kappa^2M^2 - k^2 - 4\kappa^2\phi^2}) \tag{17}$$

At the time of the start of preheating a has been taken to be 1. Damping term is neglected for the time being. ϕ decreases with time and $\langle\psi^2\rangle$ increases very fast. Therefore, within a short time growth of ψ_k and $\langle\psi^2\rangle$ gets complete. So initial form of $\langle\psi^2\rangle$ is

$$\langle\psi^2\rangle = \int_0^{\sqrt{4\kappa^2\phi^2-4\kappa^2M^2}} \frac{dk^2}{8\pi^2}\exp(2t\sqrt{4\kappa^2M^2 - k^2 - 4\kappa^2\phi^2}) \tag{18}$$

$$= \frac{1}{16\pi^2t^2}\left(1 + (2t\sqrt{4\kappa^2\phi^2 - 4\kappa^2M^2} - 1)e^{2t\sqrt{4\kappa^2\phi^2-4\kappa^2M^2}}\right) \tag{19}$$

Then after some time when $\langle\psi^2\rangle$ reaches M^2 and it settles down at that value. Growth of occupation number can also be estimated to be

$$n_k \sim |\psi_k^2| \sim \exp(2t\sqrt{4\kappa^2M^2 - k^2/a^2 - 4\kappa^2\phi^2}) \tag{20}$$

5. Discussion

For different values of κ and M different modes of ψ_k undergoes different processes. For large M like 10^{17} GeV and $\kappa = 0.01$ modes with initial physical wavenumber $k^2/a^2 = 0.3(4\kappa^2M^2)$ and $k^2/a^2 = 0.1(4\kappa^2M^2)$ undergo tachyonic growth. But for small M like 10^{15} GeV and $\kappa = 0.001$ first one of these mode undergoes parametric resonance and the second one gets tachyonic growth. A detailed discussion on this parameter dependence of the preheating process has been done in Ref.[11]. For experimental tuned values of κ and M, i.e.$M = 9.55 \times 10^{15}$GeV and $\kappa = 0.008 - 0.01$ first of the above modes shows parametric resonance and second mode grows due to tachyonic instability.

Acknowledgments

We thank Palash B Pal and Koushik Dutta for discussions and valuable suggestions.

References

[1] L. Kofman, A. D. Linde and A. A. Starobinsky, Phys. Rev. Lett. **73** (1994) 3195 [hep-th/9405187].
[2] G. R. Dvali, Q. Shafi and R. K. Schaefer, Phys. Rev. Lett. **73**, 1886 (1994) [hep-ph/9406319].
[3] S. R. Coleman and E. J. Weinberg, Phys. Rev. D **7** (1973) 1888.
[4] M. Huq, Phys. Rev. D **14**, 3548 (1976).
[5] D. Larson, J. Dunkley, G. Hinshaw, E. Komatsu, M. R. Nolta, C. L. Bennett, B. Gold and M. Halpern *et al.*, Astrophys. J. Suppl. **192**, 16 (2011) [arXiv:1001.4635 [astro-ph.CO]].
[6] N. W. McLachlan "Theory and Application of Mathieu Functions", Oxford Univ. Press, Clarendon (1947)
[7] J. Garcia-Bellido and A. D. Linde, Phys. Rev. D **57**, 6075 (1998) [hep-ph/9711360].
[8] I. Zlatev, G. Huey and P. J. Steinhardt, Phys. Rev. D **57**, 2152 (1998) [astro-ph/9709006].
[9] M. Bastero-Gil, S. F. King and J. Sanderson, Phys. Rev. D **60**, 103517 (1999) [hep-ph/9904315].
[10] G. N. Felder, L. Kofman and A. D. Linde, Phys. Rev. D **64**, 123517 (2001) [hep-th/0106179].
[11] A. Mazumdar, arXiv:1204.4113 [astro-ph.CO].

International Conference on Modern Perspectives of Cosmology and Gravitation (COSGRAV12) IOP Publishing
Journal of Physics: Conference Series **405** (2012) 012023 doi:10.1088/1742-6596/405/1/012023

Generalized Smarr formula as a local identity for arbitrary dimensional black holes

Sujoy Kumar Modak
S.N. Bose National Centre for Basic Sciences, Block-JD, Sector-III, Salt Lake City, Kolkata-700098, India
E-mail: sujoy@bose.res.in

Abstract. We discuss a method based on Killing symmetries and Komar conserved charges to generalize Smarr mass formula for arbitrary dimensional charged, rotating spacetime. We derive a local identity defined at the event horizon of the rotating black hole in Einstein-Maxwell gravity which reproduces the generalized Smarr formula as a by-product. The advantages of this new identity are the following: (i) unlike Smarr formula, which is non-local, this identity is purely local and hence a switchover between horizon and infinity is unnecessary and (ii) the new identity could be mapped with the recent investigations on emergent gravity.

1. Introduction

Smarr mass formula [1] ensures that the energy of a (3+1) dimensional charged rotating black hole is distributed as its surface energy, electrostatic energy and rotational energy: thus supports black hole *no-hair* theorem. The differential form of this formula gives the first law of black hole mechanics which is analogous to first law of thermodynamics. While this mass formula is known for few decades its extension to higher dimensions considering charged-rotating spacetimes is not known. Here we first find out to what extent the existing method [1] could be useful in order to fulfil this gap and then provide a way based on Killing symmetries and Komar conserved charges to generalize this formula to arbitrary dimensions.

This generalized formula is shown to be a by-product of a horizon identity which relates the Komar conserved charge corresponding to a null Killing vector at horizon with its thermodynamic variables. Interestingly this identity could be regarded as the analogue of another identity associated with the timelike Killing horizon of a spacetime which may not be a black hole.

2. Higher dimensional spherically symmetric black hole

In this section we use the original method of Smarr [1] to derive mass formula for $N+1$ dimensional charged black hole represented by the Reissner-Nordstrom space-time,

$$ds^2 = -(1 - \frac{m}{r^{N-2}} + \frac{q^2}{r^{2(N-2)}}) + r^2(d\theta^2 + \sin^2\theta d\phi^2)$$
$$+r^2\cos^2\theta(d\chi_1^2 + \sin^2\chi_1[d\chi_2^2 + \sin^2\chi_2(...d\chi_{N-3}^2)]). \quad (1)$$

We begin by finding the position of the event horizon by setting $g^{rr} = 0$, which is found to be $r_+ = \left(\frac{m}{2} + \sqrt{\frac{m^2}{4} - q^2}\right)^{\frac{1}{N-1}}$. Using this the area of event horizon takes the form

$$A_H = \frac{4\pi \bar{A}_{N-3}}{N-2}\left(\frac{m}{2} + \sqrt{\frac{m^2}{4} - q^2}\right)^{\frac{N-1}{N-2}} \tag{2}$$

where $\bar{A}_{N-3}$ is the area of the unit sphere in $N-3$ dimensions. By inverting this relation it is possible to express mass m as

$$m = \left(\frac{(N-2)A_H}{4\pi \bar{A}_{N-3}}\right)^{\frac{N-2}{N-1}} + q^2\left(\frac{4\pi \bar{A}_{N-3}}{(N-2)A_H}\right)^{\frac{N-2}{N-1}}. \tag{3}$$

Writing in the differential form (3) takes the form $dm = \frac{(N-2)}{(N-1)\bar{A}_{N-3}} T dA_H + \frac{2q}{(N-2)r_+^{N-2}} dq$, where, $T = \frac{\kappa}{2\pi}$ is Hawking temperature of R-N black hole. Note that in (3) we have essentially written mass in the form of a homogeneous function of degree $\frac{N-2}{N-1}$ in A_H and $q^{\frac{N-2}{N-1}}$ respectively. By virtue of this in build algebraic relationships among these parameters Euler's theorem on homogeneous function is readily applicable which finally extracts the mass (m) in terms of the variables A_H and q and the prefactors appearing in above differential form. Finally substituting m and q in terms of the physical mass (M) and charge (Q) given by $M = \frac{\bar{A}_{N-1}(N-1)}{16\pi G} m$ and $Q = \sqrt{\frac{(N-2)(N-1)\bar{A}_{N-1}}{8\pi G}}\, q$, we find [2]

$$M = Q\Phi + \frac{N-1}{N-2}\frac{\kappa A_H}{8\pi}, \tag{4}$$

where we have used the expression for electric potential $\Phi = \frac{Q}{(N-2)r_+^{N-2}}$. This is the generalized Smarr formula for the metric (1) and we see mass of the R-N black hole could be expressed in terms of electric energy and surface energy at arbitrary dimensions. For $N = 3$ it reproduces the known result.

We must also notice that the algebraic relationship among various black hole parameters plays the key role in deriving the result (4). But this is not always the case. The structure of homogeneity breaks down for rotating black holes with $N > 3$. Thus the above method does not go through and we need to find alternative way for deriving the mass formula for these cases. In the next section we highlight one such method (based on Killing symmetries and conserved charges) for $N+1$ dimensional charged, rotating black holes.

3. Higher dimensional charged, Myers-Perry black hole

The spacetime metric for the $N+1$ dimensional charged Myers-Perry black hole in Boyer-Lindquist type coordinates, with one spin parameter (a), is given by [3, 4]

$$\begin{aligned} ds^2 &= -\left(1 - \frac{m}{r^{N-4}\Sigma} + \frac{q^2}{r^{2(N-3)}\Sigma}\right)dt^2 + \frac{r^{N-2}\Sigma}{\Delta}dr^2 + \Sigma d\theta^2 - \frac{2a(mr^{N-2} - q^2)\sin^2\theta}{r^{2(N-3)}\Sigma}dtd\phi \\ &+ \left(r^2 + a^2 + \frac{a^2(mr^{N-2} - q^2)\sin^2\theta}{r^{2(N-3)}\Sigma}\right)\sin^2\theta d\phi^2 + r^2\cos^2\theta d\Omega^2_{N-3}, \end{aligned} \tag{5}$$

with the following identifications,

$$\Delta = r^{N-2}(r^2 + a^2) - mr^2 + q^2 r^{4-N}, \tag{6}$$

$$\Sigma = r^2 + a^2\cos^2\theta, \tag{7}$$

$$d\Omega^2_{N-3} = d\chi_1^2 + \sin^2\chi_1[d\chi_2^2 + \sin^2\chi_2(\cdots d\chi^2_{N-3})]. \tag{8}$$

The above metric is a solution of Einstein equation only in the slowly rotating limit (small a) and it reproduces (3) when $a = 0$. For $N = 3$ it provides the Kerr-Newman space-time. The metric (5) has two Killing vectors (timelike and spacelike) given by $\xi^\mu_{(t)} = (1, 0, 0, 0 \cdots)$ and $\xi^\mu_{(\phi)} = (0, 0, 0, 1, \cdots)$. The corresponding conserved charges is given by the Komar definition [5, 6]

$$K_{\xi^\mu_{(t)}} = -\frac{1}{8\pi}\int_{\partial\Sigma} \xi^{\mu;\nu}_{(t)} d^{N-1}\Sigma_{\mu\nu} \tag{9}$$

and

$$K_{\xi^\mu_{(\phi)}} = -\frac{1}{8\pi}\int_{\partial\Sigma} \xi^{\mu;\nu}_{(\phi)} d^{N-1}\Sigma_{\mu\nu}. \tag{10}$$

where Σ is a N dimensional spatial volume element and $\partial\Sigma$ is its boundary defined on a $N-1$ dimensional spatial hypersurface. These expressions, when evaluated at asymptotic infinity, give mass M or angular momentum $J = Ma$ up to some normalisation constant. In the next subsection we shall use these expressions and derive an identity defined at the event horizon of the space-time metric (5).

3.1. The identity $K_{\chi^\mu} = 2ST$

Black hole event horizon is a special place where the line element ds^2 vanishes. In addition, for stationary cases event horizon of a black hole is also a Killing horizon where the norm of a Killing vector ($\chi^\mu\chi_\mu$) is zero. For an axisymmetric space-time like (5) this vector turns out to be $\chi^\mu = (1, 0, 0, \Omega_H, \cdots)$ (Ω_H is the angular velocity at horizon). This can be further decomposed as $\chi^\mu = \xi^\mu_{(t)} + \Omega_H\xi^\mu_{(\phi)}$ which is nothing but a linear combination of two Killing vectors. Therefore χ^μ also satisfies the Killing equation $\nabla_\mu\chi_\nu + \nabla_\nu\chi_\mu = 0$. It could be verified that outside the horizon this vector is a non-Killing but timelike vector. It means that χ^μ is essentially a null, hypersurface orthogonal, Killing vector at the horizon. This allows us to write down a Komar conserved charge only defined at the horizon, in the following form

$$K_{\chi^\mu} = K_{\xi^\mu_{(t)}} + \Omega_H K_{\xi^\mu_{(\phi)}}. \tag{11}$$

It is possible to evaluate the integrals (9) and (10) at a finite radial distance on and outside the horizon. This constant r surface is chosen to be consists of only time synchronised events [7, 8, 9, 2]. By performing the integrals (9) and (10) we find

$$K_{\xi^\mu_{(t)}} = \frac{\bar{A}_{N-3}}{G}\left(\frac{m}{2} - \frac{q^2}{r^{N-2}} - \frac{q^2a^2(a^2+r^2)}{Nr^{N+2}}\, 2F_1(2, \frac{N}{2}; \frac{N}{2}+1; -\frac{a^2}{r^2})\right), \tag{12}$$

and

$$K_{\xi^\mu_{(\phi)}} = -\frac{a\bar{A}_{N-3}}{4Gr^{N+4}}(K_1 - K_2) \tag{13}$$

where,

$$K_1 = \frac{4mNr^{N+4} + 2q^2r^2[a^4(N-2)^2 + a^2(N-4)(N-2)r^2 - 4(N-1)r^4]}{N(N-2)} \tag{14}$$

$$K_2 = \frac{2a^2(N-2)q^2(a^2+r^2)^2\, 2F_1(1, \frac{N+2}{2}; \frac{N+4}{2}; -\frac{a^2}{r^2})}{N+2}. \tag{15}$$

Journal of Physics: Conference Series **405** (2012) 012023 doi:10.1088/1742-6596/405/1/012023

In the above expressions $2F_1(b,c;d;-\frac{a^2}{r^2}) = 1 + \frac{b.c}{1!d}(-\frac{a^2}{r^2}) + \frac{b(b+1).c(c+1)}{2!d(d+1)}(-\frac{a^2}{r^2})^2 + \cdots$ is a hypergeometric function. For $N = 3$ the above expressions reproduce the effective mass and angular momentum of Kerr-Newman black hole [9]. Finally by putting $r = r_+$ in above expressions and simplifying (11) we obtain the identity

$$K_{\chi^\mu} = 2ST, \tag{16}$$

where $S = \frac{\pi r_+^{N-3}(r_+^2+a^2)\bar{A}_{N-3}}{N-2}$ is the Bekenstein entropy and $T = \frac{(N-4)(r_+^2+a^2)+2r_+^2-(N-2)q^2 r_+^{2(3-N)}}{4\pi r_+(r_+^2+a^2)}$ is the Hawking temperature.

3.2. Generalized Smarr formula

Interestingly the above identity could be written in a slightly different form which could be identified as Smarr formula. To get that we reinstate the black hole parameters, M, J, Q, Φ, Ω_H in the left hand side of (16) whereas the right hand side is re-expressed by using $T = \frac{\kappa}{2\pi}$ and $S = \frac{A_H}{4}$. Finally for slowly rotating limit, we obtain [2]

$$M = \frac{(N-1)}{(N-2)}\Omega_H J + Q\Phi + \frac{(N-1)}{(N-2)}\frac{\kappa A_H}{8\pi} \tag{17}$$

which is the generalized Smarr mass formula for the $N+1$ dimensional charged Myers-Perry black hole.

4. Conclusion

We discussed a method to derive the generalized Smarr formula for arbitrary dimensional rotating, charged space-time in Einstein-Maxwell gravity. It was shown that the existing method based on Smarr's work [1] could not be extended beyond the spherically symmetric black holes for space-time dimensions greater than four. However, using the concepts of Killing vectors and conserved charges we were able to find the identity (16) defined at the event horizon. This identity reproduced the Smarr formula as a byproduct. By comparing the identity with the Smarr formula we note the following differences: (i) While Smarr formula (17) is dimension dependent, the identity (16) is dimension independent, (ii) while (17) is nonlocal (as some of the parameters like M, J are defined at infinity others like Φ, Ω_H etc. are defined at the horizon), (16) is completely defined at the horizon, (iii) (16) could be identified with another horizon identity $E = 2ST$ which is only defined at the Killing horizon, as proposed in [10, 11, 12] from the emergent gravity point of view. In the last case case E is the Noether charge corresponding to a diffeomorphic transformation of a metric, not associated with a black hole. From this one can speculate that for black holes E could be identified with the Komar conserve charge K_{χ^μ} as defined in (11).

References

[1] Smarr L 1973 *Phys. Rev. Lett.* **30** 71.
[2] Banerjee R, Majhi B R, Modak S Kand Samanta S 2010 *Phys. Rev.* **D 82** 124002 [arXiv:1007.5204 [gr-qc]] .
[3] Aliev A N 2006 *Phys. Rev.* **D 74** 024011.
[4] Dianyan X 1988 *Class. Quant. Grav.* **5** 871.
[5] Komar A 1959 *Phys. Rev.* **113** 934.
[6] Komar A 1962 *Phys. Rev.* **127** 1411.
[7] Cohen J M, Moses H E 1977 *Phys. Rev. Lett.* **39** 1641.
[8] Cohen J M, De Felice F 1984 *J. Math. Phys.* **25** 992.
[9] Modak S K, Samanta S 2012 *Int. J. Theor. Phys.* **51** 1416 [arXiv:1006.3445 [gr-qc]].
[10] Padmanabhan T 2004 *Class. Quan. Grav.* **21** 4485 [gr-qc/0308070].
[11] Padmanabhan T 2010 *Mod. Phys. Letts.* **A 25** 1129 [arXiv:0912.3165].
[12] Padmanabhan T 2010 *Rept. Prog. Phys.* **73** 046901 [arXiv:0911.5004 [gr-qc]].

International Conference on Modern Perspectives of Cosmology and Gravitation (COSGRAV12) IOP Publishing
Journal of Physics: Conference Series **405** (2012) 012024 doi:10.1088/1742-6596/405/1/012024

Unitary and Renormalizable Theory of Higher Derivative Gravity

Gaurav Narain[1], Ramesh Anishetty[2]

The Institute of Mathematical Sciences, Taramani, Chennai 600113, India

E-mail: gaunarain@imsc.res.in[1], ramesha@imsc.res.in[2]

Abstract. In 3+1 space-time dimensions, fourth order derivative gravity is perturbatively renormalizable. Here it is shown that it describes a unitary theory of gravitons (with/without an additional scalar) in a limited coupling parameter space which includes standard cosmology. The running of gravitational constant which includes contribution of graviton is computed. It is shown that generically Newton's constant vanishes at short distance in this perturbatively renormalizable and unitary theory.

1. Introduction and Theory

Perturbative Quantum Field Theory (QFT) has been very successful in describing three of the four known forces of nature. It has been used to construct the Standard Model of particle physics, which has been tested to a great accuracy in accelerator experiments. However when the same methods of QFT are applied to the Einstein-Hilbert (EH) theory of gravity, well known problem emerge, namely the theory is plagued with UV divergences. At each order of perturbation theory one has to add new counter terms to cancel the existing divergences, as a result the theory loses predictability. This is due to the fact that in 3+1 dimensions the Newton's constant has the mass dimension $(\text{Mass})^{-2}$. Classically EH theory is very successful in describing Cosmology [1, 2]. It was shown that when one studies quantum matter fields on curved background [3], four kind of divergences appear: $\sqrt{g}$, $\sqrt{g}R$, $\sqrt{g}R^2$, $\sqrt{g}R_{\mu\nu}R^{\mu\nu}$. This was a first hint that perhaps if when one augments the EH theory with 4-th order metric derivative terms, one might witness UV renormalizabilty. Indeed it was shown in [4] that the 4-th order metric derivative gravity is UV renormalizable using $4-\epsilon$ dimensional regularization scheme [7]. The most general action that includes all terms up to 4-th order metric derivative terms is given by,

$$A = \int \frac{\mathrm{d}^4x\sqrt{-g}}{16\pi G}\left[2\Lambda - R + \frac{\omega R^2}{6M^2} - \frac{R_{\mu\nu}R^{\mu\nu} - \frac{1}{3}R^2}{M^2}\right], \tag{1}$$

where R is the Ricci scalar and $R_{\mu\nu}$ is the Ricci tensor. The last term is proportional to the square of Weyl tensor in 3+1 dimensions up to an Euler characteristic. Here ω is dimensionless and M has dimensions of mass.

Doing perturbations around the flat background $g_{\mu\nu} = \eta_{\mu\nu} + h_{\mu\nu}$ ($\eta_{\mu\nu} = \{1, -1, -1, -1\}$) one obtains the propagator of the theory [5, 6]. Here we will consider the action A for $\Lambda = 0$ in the Landau gauge $\partial^\mu h_{\rho\mu} = 0$, where the Feynman propagator of theory in momentum space is given

Journal of Physics: Conference Series **405** (2012) 012024 doi:10.1088/1742-6596/405/1/012024

by,

$$D_{\mu\nu,\alpha\beta} = \frac{i\,16\pi G}{(2\pi)^4} \cdot \left[\frac{(2P_2 - P_s)_{\mu\nu,\alpha\beta}}{q^2 + i\,\epsilon} + \frac{(P_s)_{\mu\nu,\alpha\beta}}{q^2 - M^2/\omega + i\epsilon} - \frac{2\,(P_2)_{\mu\nu,\alpha\beta}}{q^2 - M^2 + i\epsilon} \right], \tag{2}$$

where q is the momentum of fluctuating field $h_{\mu\nu}$. Various spin projectors are $(P_2)_{\mu\nu,\alpha\beta} = \frac{1}{2}\left[T_{\mu\alpha}T_{\nu\beta} + T_{\mu\beta}T_{\nu\alpha}\right] - \frac{1}{3}T_{\mu\nu}T_{\alpha\beta}$, $(P_s)_{\mu\nu,\alpha\beta} = \frac{1}{3}T_{\mu\nu}T_{\alpha\beta}$, where $T_{\mu\nu} = \eta_{\mu\nu} - q_\mu q_\nu/q^2$. The first term of Eq. (2) is the massless spin-2 graviton with two degrees of freedom, the second term is the scalar (Riccion) of mass $M/\sqrt{\omega}$ and the last term is that of massive spin-2 with mass M (M-mode). The last term arises due to the presence of $R_{\mu\nu}R^{\mu\nu}$ term in the action. If we only had $F(R)$ type of theory, this term will be absent. From the propagator Eq. (2) we note that the residues at the pole for graviton and Riccion are positive, while for the M-mode it is negative. The M-mode renders the theory non-unitary.

We now study the issue of unitarity of higher derivative gravity action given in Eq. (1). The full S-matrix of theory involve gravitons, Riccions and M-mode as the external legs. Lets consider a subpart of this S-matrix which involve only Riccion and graviton as external legs, which means that we are considering scattering process which involve only Riccion and graviton but no M-mode as external leg. In this subpart of S-matrix we ask whether this under some conditions remains unitary? In any such scattering process, M-mode can appear as an intermediate state, which is not present in our subpart of S-matrix. For intermediate energies μ less than M it cannot occur anyway. In renormalized quantum field theory these mass parameter are also energy dependent. Hence if $M(\mu)$ is always greater than μ then it cannot occur at any intermediate energy, consequently the subpart of S-matrix can also satisfy unitarity. In this scenario the physical amplitudes do get contributions from the M-mode but not from the imaginary part of M-mode *i.e.* Cutkosky cut corresponding M-mode vanish identically. Thus in the renormalized theory if $M^2(\mu)/\mu^2 > 1$ is satisfied, then we can have unitary theory of only gravitons and Riccions [9]. In the following we will see how this can be realized.

2. Renormalization Group Analysis

We choose to parameterize our gravity action as in Eq. (1), so that in the path-integral G effectively plays the same role as $\hbar$ (in the absence of matter fields) *i.e.* the loop expansion is same as perturbation theory in small G. To one-loop beta function of couplings have been computed using the Schwinger-Dewitt technique [8]. In [10, 11] this was done using dimensional regularization in $4-\epsilon$ space-time dimensions. In the Landau gauge the beta function are the following [12]:

$$\frac{\mathrm{d}}{\mathrm{d}t}\left(\frac{1}{M^2 G}\right) = -\frac{133}{10\pi}, \tag{3}$$

$$\frac{\mathrm{d}}{\mathrm{d}t}\left(\frac{\omega}{M^2 G}\right) = \frac{5}{3\pi}\left(\omega^2 + 3\omega + \frac{1}{2}\right), \tag{4}$$

$$\frac{\mathrm{d}}{\mathrm{d}t}\left(\frac{1}{G}\right) = \frac{5M^2}{3\pi}\left(\omega - \frac{7}{40\omega}\right), \tag{5}$$

$$\frac{\mathrm{d}}{\mathrm{d}t}\left(\frac{2\Lambda}{G}\right) = \frac{M^4}{2\pi}\left(5 + \frac{1}{\omega^2}\right) - \frac{4M^2\Lambda}{3\pi}\left(14 + \frac{1}{\omega}\right), \tag{6}$$

where $t = \ln(\mu/\mu_0)$ and all rhs contains the leading contribution in G (M^2G is also taken to be small), with higher powers coming from higher loops being neglected. Using Eqs. (3 and 4) we solve for the running of ω,

$$\frac{\mathrm{d}\omega}{\mathrm{d}t} = \frac{5M^2G}{3\pi}\left(\omega^2 + \frac{549}{50}\omega + \frac{1}{2}\right) = \frac{5M^2G}{3\pi}(\omega + \omega_1)(\omega + \omega_2), \tag{7}$$

where $\omega_1 = 0.0457$ and $\omega_2 = 10.9343$. On studying Eq. (7) we find that the RG flow of ω has two fixed point: $-\omega_1$ and $-\omega_2$, with former being repulsive and later attractive under the UV evolution or increasing t. We realize that both these fixed points lie in the unphysical domain. The last equality of Eq. (7) tells that rhs is always positive for $\omega > 0$. This means that ω is a monotonic increasing function of t and vice-versa. Eq. (3) is easily integrated to express the flow of M^2G in terms of t. This is then plugged in Eq. (7) to obtain,

$$t = T\left[1 - \left(\frac{\omega+\omega_2}{\omega+\omega_1}\cdot\frac{\omega_0+\omega_1}{\omega_0+\omega_2}\right)^\alpha\right], \tag{8}$$

where $T = 10\pi/(133M_0^2G_0)$ and $\alpha = 399/50(\omega_2 - \omega_1)$, with subscript 0 meaning that the coupling parameters are evaluated at $t = 0$ or $\mu = \mu_0$. As ω takes value between zero and infinity, this translates using Eq. (8) in to a range for t. We note that t takes a minimum value for $\omega = 0$ while a maximum value for $\omega = \infty$.

$$\frac{t_{min}}{T} \equiv 1 - \left(\frac{\omega_2}{\omega_1}\frac{\omega_0+\omega_1}{\omega_0+\omega_2}\right)^\alpha \leq \frac{t}{T} < 1 - \left(\frac{\omega_0+\omega_1}{\omega_0+\omega_2}\right)^\alpha \equiv \frac{t_{max}}{T}. \tag{9}$$

Using the evolution equation for ω, one can transform any evolution of coupling in t space to ω space. This allows analytical expressions for the flow of other couplings. Using Eqs. (5 and 7) we get,

$$\frac{\mathrm{d}\ln G}{\mathrm{d}\omega} = -\frac{\omega - \frac{7}{40\omega}}{(\omega+\omega_1)(\omega+\omega_2)}, \quad \frac{G}{G_0} = \frac{\omega_0}{\omega}\cdot\left(\frac{1+\omega_1/\omega}{1+\omega_1/\omega_0}\right)^{A1}\left(\frac{1+\omega_2/\omega}{1+\omega_2/\omega_0}\right)^{A2}, \tag{10}$$

where the first equation expresses the running of $\ln G$ in ω space, while in second equation we write its solution. The first equation tells that during the RG evolution G gets extremised at $\omega = \sqrt{7/40}$, taking the second derivative it is shown that this point is a maxima. We choose this point to be our reference point μ_0 or $t = 0$ and integrate to obtain the second equation, where $A1 = -0.3473$ and $A2 = -1.0027$. Eq. (10) show that for large ω or t, $G \sim 1/\omega$, thereby vanishing for large t [16], while for small ω, $G \sim \omega^{7/20}$ reaching a peak at $\omega_0 = \sqrt{7/40}$. Similarly, using Eqs. (4 and 5) we obtain the running of M^2/ω, which along with with Eq. (7) is integrated to give $M^2/\omega = (M_0^2/\omega_0)\left((1+\omega_1/\omega)/(1+\omega_1/\omega_0)\right)^{B1}\left((1+\omega_2/\omega)(1+\omega_2/\omega_0)\right)^{B2}$, where $B1 = 1.0802$ and $B2 = 0.2698$. This tells that as $\omega \to \infty$, the mass of the M-mode also goes to infinity, which means that ultimately it is decoupled from the theory. But does it goes to infinity quick enough so that it is never encountered in the theory, is the question we ask next? To answer this we now analyze M^2/μ^2. It is instructive to note that from Eqs. (3, 5 and 7) we obtain the running of $\ln(M^2/\mu^2)$,

$$\frac{\mathrm{d}}{\mathrm{d}\omega}\ln\left(\frac{M^2}{\mu^2}\right) = \frac{\left(\omega + \frac{399}{50} - \frac{7}{40\omega} - \frac{6\pi}{5M^2G}\right)}{(\omega+\omega_1)(\omega+\omega_2)}. \tag{11}$$

This shows that M^2/μ^2 reaches a minima for $\omega = \omega_*$ given by $(\omega_* + 399/50 - 7/40\omega_*) = 6\pi/(5M_*^2G_*)$. Hence by demanding $M_*^2/\mu_*^2 = (6\pi/5\mu_*^2G_*)/(\omega_* + 399/50 - 7/40\omega_*) > 1$, we make the M-mode not realizable in the physical GR sector of the theory. This inequality is easily achievable by choosing $\mu_*^2G_*$ appropriately. Perturbative loop expansion requires that M^2G is small. Therefore M is a sub-Planckian mass, yet the running mass as dictated by interactions makes it physically not realizable even in post Planckian regime.

One can do a similar analysis to answer the question whether Riccion is realizable or not? For this we study the RG evolution of $M^2/(\omega\mu^2)$. From Eqs. (4, 5 and 7) we obtain $\mathrm{d}\ln\left(M^2/\omega\mu^2\right)/\mathrm{d}\omega = -\left(3 + 27/40\omega + 6\pi/5M^2G\right)/(\omega+\omega_1)(\omega+\omega_2)$, showing that the Riccion mass relative to μ decreases monotonically. By a suitable choice we can make the Riccion to be physically realizable or not. So we conclude that there exists unitary physical subspace only with the gravitons or along with Riccions.

To make the running of Λ to zero we add two spin-$\frac{1}{2}$ Dirac fields (for detail see [9]) with mass $(5/4)^{1/4}M$ and $M/\sqrt{2\omega}$ [13] (these additional fermionic fields can be interpreted as ghosts normalizing the functional integral). This assures that if initially $\Lambda = 0$, then throughout the RG flow Λ will remain zero. The affect of adding the two Dirac fields is just to shift ω_0, ω_1, ω_2 and ω_*, but all conclusions remains unaltered.

In arbitrary gauge the beta function $1/M^2G$ and ω/M^2G remains unaltered, which means that running of ω and two fixed points ω_1 and ω_2 remains same. However the beta function of $1/G$ gets modified. In a general gauge it is $\mathrm{d}\ln G/\mathrm{d}t = -(5M^2G/3\pi)(\omega + a + b/\omega)$, where a and b depend on the gauge-fixing parameters used, while the leading term is gauge invariant. This means that ω_0 shifts. Doing the same analysis as before, we write the running of G in terms of ω, from which we note that for large ω or t, $G \sim 1/\omega$ is a gauge invariant result, while for small ω, the rate at which G vanishes depend on a and b.

3. Discussion

By doing the one-loop analysis we have found that there is no solution for ω from Eq. (8) if $t > t_{max}$. This is because ω reaches its maximum value infinity. In the other extreme when $\omega = 0$, t reaches its minimum value t_{min}. For $t < t_{min}$, ω become negative *i.e.* M^2/ω the Riccion mass square becomes negative signaling the instability of the vacuum.

Finally we conclude that the action A, Eq. (1) describes a perturbative quantum gravity as self consistent, renormalizable and unitary theory of gravitons and the curvature cannot become singular, in particular it cannot fluctuate wildly at sub-Planckian [15] or post Planckian regimes consistent with known cosmology. Its predictions asymptotically beyond Planck scale needs to be investigated further.

References

[1] Dodelson S, "Modern Cosmology," Amsterdam, Netherlands: Academic Press, (2003) 440 p.
[2] Weinberg S, "Cosmology," Oxford, UK: Oxford Univ. Press, (2008) 593 p.
[3] Utiyama R and DeWitt B S, J. Math. Phys. **3**, 608-618 (1962).
[4] Stelle K S, Phys. Rev. D **16** (1977) 953;
[5] Salam A and Strathdee J A, Phys. Rev. D **18** (1978) 4480.
[6] Julve J and Tonin M, Nuovo Cim. B **46** (1978) 137.
[7] 't Hooft G and Veltman M J G, NATO Adv. Study Inst. Ser. B Phys. **4** (1974) 177-322.
[8] DeWitt B S, Phys. Rev. Lett. **13**, 114-118 (1964).
[9] Narain G and Anishetty R, Phys. Lett. B **711**, 128 (2012) [arXiv:1109.3981 [hep-th]].
[10] Fradkin E S and Tseytlin A A, Phys. Lett. B **104** (1981) 377.
[11] Avramidi I G and Barvinsky A O, Phys. Lett. **B159** (1985) 269-274.
[12] Avramidi I G, "Heat kernel and quantum gravity," Lect. Notes Phys. **M64** (2000) 1-149.
[13] Gorbar E V and Shapiro I L, JHEP **0306** (2003) 004. [hep-ph/0303124].
[14] 't Hooft G and Veltman M J G, Annales Poincare Phys. Theor. A **20** (1974) 69.
[15] Starobinsky A A, Phys. Lett. B **91** (1980) 99.
[16] G is indeed a nontrivial function of $\ln \mu/\mu_0 = t$ through $\omega(t)$, consequently the criticisms alluded in " Anber M M and Donoghue J F, arXiv:1111.2875 [hep-th]." do not pertain to our analysis.

Journal of Physics: Conference Series **405** (2012) 012025 doi:10.1088/1742-6596/405/1/012025

Possible role of Berry phase in inflationary cosmological perturbations

Barun Kumar Pal, Supratik Pal and B. Basu

Physics and Applied Mathematics Unit, Indian Statistical Institute, 203 B.T. Road, Kolkata 700 108, India

E-mail: barunp1985@rediffmail.com, supratik@isical.ac.in, banasri@isical.ac.in

Abstract. Here we have derived a cosmological analogue of Berry phase by obtaining the corresponding wavefunction for the system of inflationary cosmological perturbations solving the Schrodinger equation. We have further shown that cosmological Berry phase can be related inflationary observable parameters. As a result one can, atleast in principle, establish a supplementary probe of inflationary cosmology through the measurement of the associated Berry phase. But we do not make any strong comment on this.

1. Introduction

The original motivation behind inflation [1] was to resolve the puzzles associated with *Standard Big Bang* scenario but the most exciting aspect of this scenario comes from the fact that they provide natural explanation for the quantum origin of classical cosmological fluctuations [2] observed in the large scale structure of the matter and in the cosmic microwave background. The accelerated expansion converts the initial vacuum quantum fluctuations into macroscopic cosmological perturbations. So, measurement of any quantum property which reflects on classical observables serve as a supplementary probe of inflationary cosmology, complementing the well-known CMB polarization measurements [3]. This has led us to investigate for the potentiality of Berry phase [4] in providing a measurable quantum property which is inherent in the macroscopic character of classical cosmological perturbations, thereby serving as a supplementary probe to CMB in inflationary cosmology.

Here our intention is to demonstrate the effect of the curved spacetime background in the dynamical evolution of the quantum fluctuations during inflation through the derivation of the associated Berry phase and search for the possible consequences via observable parameters. To this end, we first find an exact wavefunction for the system of inflationary cosmological perturbation by solving the associated Schrodinger equation. The relation between the dynamical invariant [5, 6, 7] and the geometric phase has then been utilized to derive the corresponding Berry phase [8, 9]. For slow roll inflation the total accumulated phase gained by each of the modes during sub-Hubble oscillations (adiabatic limit) is found to be a new parameter made of corresponding (scalar and tensor) spectral indices. Since tensor spectral index is related to the tensor to scalar amplitude ratio through the consistency relation, the Berry phase can indeed be utilized to act as a supplementary probe of inflationary cosmology.

2. Inflationary Cosmological perturbations as parametric oscillators

The total action for the system of single scalar field driven inflation is given by

$$S = -\frac{1}{2}\int d^4x\sqrt{-g}\left[M_P^2 R + g^{\mu\nu}\phi_{,\mu}\phi_{,\nu} + 2V(\phi)\right] \tag{1}$$

where R is the Ricci scalar, ϕ the inflaton and $V(\phi)$ being corresponding potential. The explicit expression for the perturbations quantities can be calculated by varying the above action. The single scalar field driven inflation produces both the *scalar* and *tensor* perturbations, but in the linear theory they evolve independently as a result we can study their evolution separately. It can be shown that for the scalar perturbation everything can be reduced to the study of a single variable v known as Mukhanov-Sasaki variable defined by, $v \equiv a\delta\phi_{GI} + z\Phi_B$ with $z = a\frac{\phi'}{\mathcal{H}}$, $\delta\phi_{GI}$ gauge invariant inflaton fluctuation and Φ_B Bardeen's Potential [10]. For the tensor perturbation it is the variable u defined by, $h_{ij} = \frac{u}{a}Q_{ij}$ where Q_{ij} Eigentensors of the Laplacian operator.

The actions for the scalar and tensor perturbations can be cast into the following form [11]

$$\begin{aligned}\mathbf{S}^S &= \frac{1}{2}\int d\eta d\mathbf{x}\left[\Pi^2 + \frac{z'}{z}(\Pi v - v\Pi) - \delta^{ij}\partial_i v\partial_j v\right], \text{ where } \Pi \equiv \frac{\partial \mathbf{L}}{\partial v'} = v' - \frac{z'}{z}v \\ \mathbf{S}^T &= \frac{1}{2}\int d\eta d\mathbf{x}\left[\pi^2 + \frac{a'}{a}(\pi u - u\pi) - \delta^{ij}\partial_i u\partial_j u\right], \text{ where } \pi \equiv \frac{\partial \mathbf{L}}{\partial u'} = u' - \frac{a'}{a}u\end{aligned} \tag{2}$$

The variable v is related to the most fundamental perturbation quantity, comoving curvature perturbation $\mathcal{R}$, via the relation $v = -z\mathcal{R}$. Promoting the fields to operators and taking the Fourier decomposition, the Hamiltonian operators corresponding to the above actions (2) is found to be

$$\hat{\mathbf{H}}_{\mathbf{k}} \equiv \sum_{j=1}^{2}\hat{\mathbf{H}}_{j\mathbf{k}} = \sum_{j=1}^{2}\frac{1}{2}\left[k^2\hat{q}_{j\mathbf{k}}^2 + Y(\eta)\left(\hat{p}_{j\mathbf{k}}\hat{q}_{j\mathbf{k}} + \hat{q}_{j\mathbf{k}}\hat{p}_{j\mathbf{k}}\right) + \hat{p}_{j\mathbf{k}}^2\right] \tag{3}$$

here we have decomposed $v_{\mathbf{k}}$, $u_{\mathbf{k}}$, $\Pi_{\mathbf{k}}$, $\pi_{\mathbf{k}}$ into their real and imaginary parts. Also $\hat{q}_{j\mathbf{k}} = \hat{v}_{j\mathbf{k}}$, $\hat{u}_{j\mathbf{k}}$; $\hat{p}_{j\mathbf{k}} = \hat{\Pi}_{j\mathbf{k}}$, $\hat{\pi}_{j\mathbf{k}}$ and $Y = \frac{z'}{z}$, $\frac{a'}{a}$ for the scalar and tensor modes respectively and $j = 1,2$ with the frequency given by $\omega = \sqrt{k^2 - Y^2}$. So for each of the scalar and tensor modes, the Hamiltonian is a sum of two parametric oscillators, each of them having the form (3).

3. The Berry phase in inflationary cosmological perturbations

To calculate the Berry phase we require the wavefunction of the system. To this end we first solve the associated Schrdinger equation using the Dynamical Invariant Operator Method (DIOM) [7] for the system of inflationary cosmological perturbations. Following the usual technique [5, 6], after some straightforward but tedious algebra we found

$$\Psi_{n_1,n_2} = \frac{e^{i\alpha_{n_1,n_2}(\eta)}\bar{H}_{n_1}\left[\frac{q_{1\mathbf{k}}}{\rho_k}\right]\bar{H}_{n_2}\left[\frac{q_{2\mathbf{k}}}{\rho_k}\right]}{\sqrt[4]{\pi^2 2^{2(n_1+n_2)}(n_1!n_2!)^2\rho_k^4}}\exp\left[\frac{i}{2}\left(\frac{\rho_k'}{\rho_k} - Y(\eta) + \frac{i}{\rho_k^2}\right)\left(q_{1\mathbf{k}}^2 + q_{2\mathbf{k}}^2\right)\right] \tag{4}$$

where ρ_k is a time dependent real function satisfying the Milne-Pinney equation

$$\rho_k'' + \Omega^2(\eta,k)\rho_k = \frac{1}{\rho_k^3(\eta)}, \text{ where } \Omega^2 = \omega^2 - \frac{dY}{d\eta}. \tag{5}$$

Once the wavefunction of the system is known it can then be used to derive the associated Berry phase and in the present context we have found

$$\gamma_{n_1,n_2,k} = -\frac{1}{2}(n_1+n_2+1)\int_0^{\Gamma}\left(\frac{1}{\rho_k^2}-\rho_k^2\omega^2-(\rho_k')^2\right)d\eta \tag{6}$$

where it has been assumed that the invariant is Γ periodic and its eigenvalues are non-degenerate. Now in the adiabatic limit Eqn.(5) can be solved [6] by a series of powers in adiabatic parameter, so at the ground state of the system the Berry phase for a particular perturbation mode can be evaluated upto the first order in adiabatic parameter and is given by

$$\gamma_k^{(S,\,T)} = \frac{1}{2}\int_0^{\Gamma}\frac{Y'}{\sqrt{k^2-Y^2}}d\eta \tag{7}$$

where the superscripts S and T stand for scalar and tensor modes respectively.

4. Berry Phase interms of cosmological observables

Now to link this Berry phase with the cosmological observables we need to fix the value of the parameter Γ. To this end we shall calculate the total Berry phase accumulated by each mode during inflationary sub-Hubble oscillations which is

$$\gamma_{k\ sub}^{S,T} = -\frac{1}{2}\lim_{\eta'\to-\infty}\int_{\eta'}^{\eta_0^{S,T}}\frac{Y'}{\sqrt{k^2-Y^2}}d\eta \tag{8}$$

where $\eta_0^{S,T}$ is the conformal time which satisfies the relation $k^2=\left[Y(\eta_0^{S,T})\right]^2$ and guarantees that the modes are within the horizon and oscillating with real frequency.

The *accumulated Berry phase* during sub-Hubble evolution of the scalar modes, in terms of the slow-roll parameters, turns out to be

$$\gamma_k^S = \frac{1}{2}\lim_{\eta'\to-\infty}\int_{\eta'}^{-\frac{\sqrt{1+6\epsilon_1-2\epsilon_2}}{k}}\frac{\frac{z''}{z}-\left(\frac{z'}{z}\right)^2}{\sqrt{k^2-\left(\frac{z'}{z}\right)^2}}d\eta \approx -\frac{\pi}{4}\frac{1+3\epsilon_1-\epsilon_2}{\sqrt{1+6\epsilon_1-2\epsilon_2}}+O(\epsilon_1^2,\epsilon_2^2,\epsilon_1\epsilon_1) \tag{9}$$

Similarly for the tensor modes we have

$$\gamma_k^T = \frac{1}{2}\lim_{\eta'\to-\infty}\int_{\eta'}^{-\frac{\sqrt{1+2\epsilon_1}}{k}}\frac{\frac{a''}{a}-\left(\frac{a'}{a}\right)^2}{\sqrt{k^2-\left(\frac{a'}{a}\right)^2}}d\eta \approx -\frac{\pi}{4}\frac{1+\epsilon_1}{\sqrt{1+2\epsilon_1}}+O(\epsilon_1^2,\epsilon_2^2,\epsilon_1\epsilon_1) \tag{10}$$

where ϵ_1, ϵ_2 are the usual potential slow-roll parameters. As, during inflation the slow-roll parameters do not evolve significantly from their initial values so in the above estimates for $\gamma_k^{S,T}$ we can consider ϵ_1 and ϵ_2 as their values at horizon crossing without committing any substantial error. Also at the horizon exit the fundamental observable parameters can be expressed in terms of slow-roll parameters, so we found the *accumulated Berry phase* during inflationary *sub-Hubble* oscillations of the modes to be given by

$$\gamma_k^S \approx -\frac{\pi}{8}\frac{3-n_S(k)}{\sqrt{2-n_S(k)}},\quad n_S(k)\approx 3-\frac{8\gamma_k^S}{\pi}\left(\frac{4\gamma_k^S}{\pi}-\sqrt{\frac{16[\gamma_k^S]^2}{\pi^2}-1}\right) \tag{11}$$

$$\gamma_k^T \approx -\frac{\pi}{8}\frac{2-n_T(k)}{\sqrt{1-n_T(k)}},\quad n_T(k)\approx 2-\frac{8\gamma_k^T}{\pi}\left(\frac{4\gamma_k^T}{\pi}-\sqrt{\frac{16[\gamma_k^T]^2}{\pi^2}-1}\right) \tag{12}$$

where n_S and n_T are the scalar and tensor spectral indices respectively. So the Berry phase for sub-Hubble oscillations of the perturbation modes during inflation can be completely envisioned through the observable parameters.

5. Discussion and Summary

In this article we have demonstrated how the exact expression for the wave function of the quantum cosmological perturbations can be analytically obtained by solving the associated Schrodinger equation following the dynamical invariant technique. This helps us to derive an expression for cosmological analogue of Berry phase. The classical cosmological perturbation modes (both scalar and tensor) having quantum origin picks up a phase during their advancement through the curved space-time background that depends entirely on the background geometry and can be estimated quantitatively by measuring the corresponding spectral indices. So the Berry phase for the quantum counterpart of the classical cosmological perturbations endow us with the measure of spectral index.

Any attempt towards the measurement of cosmological Berry phase may reflect observational credentials of this parameter in inflationary cosmology. So, cosmological Berry phase may have the potentiality to play some important role in inflationary cosmology. However, we are yet to figure out this quantitative feature in a more concrete language. So, we do not make any strong comment on this. At this point, all we would like to point out is that an analogue of Berry phase appears in inflationary cosmology, which is a measurable observable of the quantum property of the cosmological perturbations and which may presumably give some insight on observable parameters as well.

So far as the detection of cosmological Berry phase is concerned, we are far away from quantitative measurements. A possible theoretical aspect of detection [12] of the analogue of cosmological Berry phase may be developed in squeezed state formalism [11]. For a quantum harmonic oscillator, when a squeezing Hamiltonian is switched on, and the squeezing parameter is varied, we can find a detectable Berry phase. As the inflationary perturbations can be studied in the squeezed state formalism, we hope to put forward our analysis on the detection of the geometric phase in near future.

Acknowledgments

BKP thanks Council of Scientific and Industrial Research, India for financial support through Senior Research Fellowship (Grant No. 09/093 (0119)/2009).

References

[1] A. H. Guth, Phys. Rev. D **23**, 347 (1981)
[2] A. H. Guth and S. Y. Pi, Phys. Rev. Lett. **49**, 1110 (1982); V. Mukhanov et al., Phys. Rep. **215**, 203 (1992)
[3] http://lambda.gsfc.nasa.gov/product/map/current
[4] M. V. Berry, Proc. Roy. Soc. A**392**, 45 (1984)
[5] H. R. Lewis, Jr., Phys. Rev. Lett. **18**, 636 (1967)
[6] H. R. Lewis, Jr., J. Math. Phys. **9**, 1997 (1968)
[7] H. R. Lewis, Jr. and W. B. Riesenfeld, J. Math. Phys. **10**, 1458 (1969)
[8] D. A. Morales, J. Phys. A: Math. Gen. **21**, L889 (1988)
[9] D. B. Monteoliva et al., J. Phys. A: Math. Gen. **27**, 6897 (1994)
[10] J. M. Bardeen, Phys. Rev. D **22**, 1882 (1980)
[11] A. Albrecht et. al, Phys. Rev. D **50**, 4807 (1994)
[12] I. Fuentes-Guridi et al., Phys. Rev. Lett. **85** 5081 (2000)

Journal of Physics: Conference Series **405** (2012) 012026 doi:10.1088/1742-6596/405/1/012026

Cluster merger blast wave and the mystery of ringlike radio-relic formation around some galaxy clusters

Surajit Paul

Inter University Centre for Astronomy and Astrophysics, Post Bag 4, Ganeshkhind, Pune University Campus, Pune 411007, India

E-mail: surajit@iucaa.ernet.in

Abstract. In this work I studied the nature and important effects of massive galaxy cluster merger phenomena. Due to inherent complexity of such events analytical solution is impossible, so, numerical simulations are performed using ENZO-2.1 hydrodynamic code. It is noticed that the formation of Mega parsec scale merger shocks in such events substantially change the energy distribution of Inter Cluster Medium. A striking similarity is noticed between expanding intra cluster medium during mergers with the blast wave formation in supernovae explosion. The blast wave meets the void/ accretion shocks when propagated out to the virial radius. Particle acceleration at the meeting point produce a significant amount of synchrotron radio emission through which curved shocks are made visible in radio waves. This study thus also sheds some light on the formation of curved and nearly symmetric radio emission found in Abell 3376, Abell 3667, CIZA J2242.8+5301, plck g287.0+32.9 etc. clusters.

1. Introduction

Remarkable technical development in the field of radio interferometry revealed many spectacular structures in the recent years. Among them the most mysterious are the symmetric Mpc scale radio relics around massive galaxy clusters (e.g. Abell 3667, Abell 3376 etc.) found near or beyond the virial radius (e.g.[1]). Origin of these peripheral ring like radio formation couldn't be explained properly and remained an outstanding problem for the last decade.

In the hierarchical structure formation framework, clusters are the largest objects in the universe that may have attained virialisation recently, and at present many clusters are still in the process of growing by accretion and mergers. When the mass ratio of the infalling cluster halos in a merger ($> 10^{13} M_{\odot}$) approaches unity, a gigantic clash takes place. These specific merger events, termed as 'major mergers' [6], are among the most energetic events of the universe and energy released in these events may approach 10^{64} ergs.

In a major merger, sudden increase in bulk motion of ICM produces Mpc-scale shock fronts. Strong collisionless shocks are capable of producing high-energy cosmic-ray particles (CR) via diffusive shock acceleration mechanism (DSA; [2]). These curved shocks are also interesting for transforming kinetic energy to turbulent energy by injection of volume-filling turbulence in the ICM [3]. On the other hand, if a turbulent flow is established in the ICM, turbulent dissipation acts on a significantly longer timescale than shocks, and in principle can stochastically re-accelerate the ambient electrons [4] and also amplify magnetic fields by shock compression and dynamo action. Cluster mergers thus greatly affects the energy budget of the ICM and can raise turbulent to thermal energy fraction to as high as 30% ([5], [6] etc.).

2. Simulation of galaxy-cluster mergers

Due to the inherent complexity of the merger events, hydro-dynamical simulations became an important tool for the study of hierarchical assembly of large structures and their evolution. With an aim to understand cluster merger phenomena well, I performed several simulations of major cluster mergers using the ENZO 2.1 hydro-dynamical code [7].

Simulations were performed with the Adaptive Mesh Refinement (AMR), grid-based hybrid (N-body plus hydro-dynamical) code Enzo v. 2.1 [7]. A flat ΛCDM background cosmology with the parameters of the LCDM model, derived from WMAP (5-years data) combined with the distance measurements from the Type Ia supernovae (SN) and the Baryon Acoustic Oscillations (BAO) is used (see [8]). The simulations have been initialized at redshift $z = 60$ using the [9] transfer function, and evolved up to $z = 0$. An ideal equation of state was used for the gas, with $\gamma = 5/3$. Radiative cooling assumed from [10] for a fully ionized gas with metallicity of 0.5 solar.

3. Emergence and evolution of supernova type blast wave

Unlike many previous attempt of controlled binary mergers ([11]) where binary bow type shocks are found, in this realistic cosmic environment simulations a supernovae type blast wave is created after the massive mergers (see Fig 1). Like a blast wave, here, cluster medium pressure expanding supersonically outward from the merging core. This enormous energy comes from the binding energy of new core formation. It has a leading shock front of compressed gases called a merger shock front. It is also noticed that once the shock is emerged, the central dark matter potential has no further control on the evolving shock fronts. Its kinetic energy of wave propagation is not exactly spherically symmetric like most of the blast waves, rather, most of the energy here transferred along the merging axis making it an ellipsoidal shock front [6].

Three panels of the Fig 1 shows the evolution of temperature (pseudo colour) and density (contour) as a time series starting from pre-merger to post merger condition. A clear launch of thermal shock (2nd panel) can be seen after the merger. A ring like formation of shock wave strikingly matches the morphology of a blast wave propagation. So, this is possibly the biggest known blast wave in this universe with a size of > 1 Mpc.

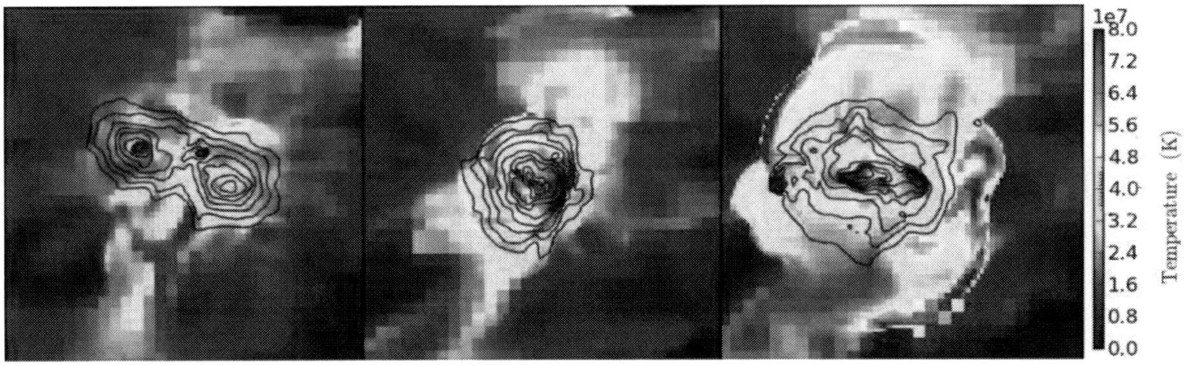

Figure 1. Pseudo colour temperature evolution map with density contours are shown for redshift z=0.88, z=0.76 and z=0.68. Slices are cut at the centre of mass with each side of 3 Mpc.

3.1. Expansion of the cluster medium

Cluster merger blast wave after emergence, propagate in to the cluster medium. High pressure that was created due to the release of cluster core binding energy then transferred to the cluster medium through shocks. Materials from the cluster centre then expanded radially outwards against the core collapse. In the Fig 2 this process is clearly depicted by velocity vector image. First panel in the figure 2 shows the Pre-merger condition. At this stage two clusters cores are about to merge and the gravitational core collapse and accretion flow is looking dominant force. Materials are falling in to the cluster core which is indicated by the inward vector flow direction. Next panel shows the emergence of a shock and materials started flowing outwards. Velocity vector inside the shock from now turned around radially outwards. Only matter along the filaments are still flowing inside. In the last panel it is clearly seen that a ring like shock front is created and material from the core of the cluster radially pushed towards the virial radius of the cluster. An almost hard surface is created at the shock front where velocity vectors are seen to reverse their direction (See panel 3 Fig 2). This ellipsoidal formation got brocken where ever materials are flowing towards the core through filaments.

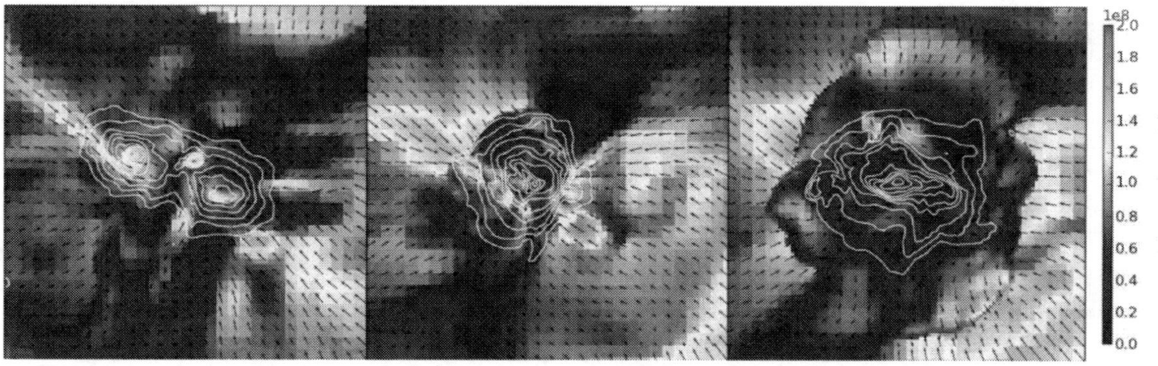

Figure 2. Same as Fig 1 with density as white contours (1×10^{-28} *to* $1 \times 10^{-24} gmcm^{-3}$, velocity magnitude in pseudo clolour and flow direction as velocity vectors.

3.2. Blast wave energy evolution

Radial variation of total energy of the cluster system is computed. Here total energy means the addition of thermal and the kinetic energy of the cluster medium. Snapshots are taken at

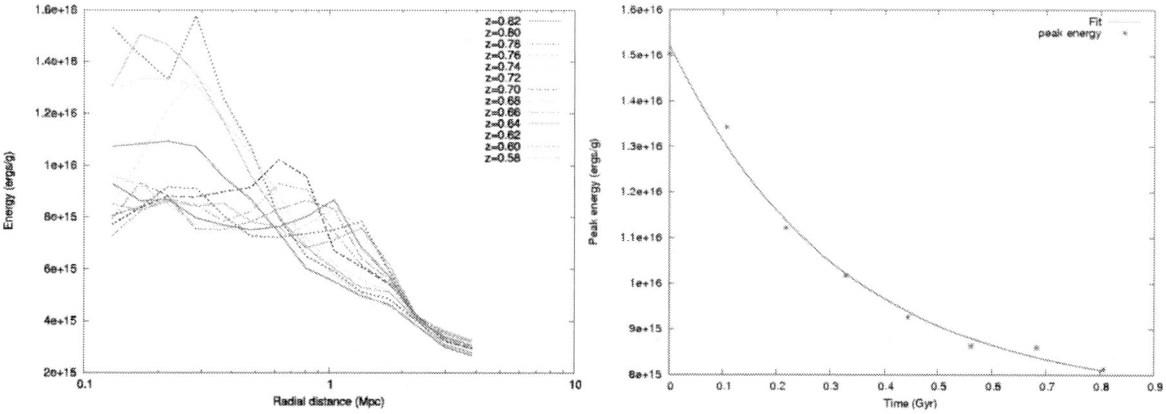

Figure 3. Panel 1: Radial total energy evolution for each snapshots starting from redshift z=0.82 to z=0.58. **Panel 2:** Peak energy evolution as points and fitting as line plot

each 0.02 redshift separation and plotted the full evolution of the cluster from redshift z=0.82

upto redshift z=0.58. This cluster merged almost 6.787 Giga year before and blast wave can be traced for more than a Giga year. The radial profile of energy evolution shows almost spheroidal type of energy increment along the radius. It shows an initial increment in the peak energy at the shock front and slowly it looses its energy as goes outwards Fig 3. This profile is exactly similar (with some scaling) to a supernova blast wave. The peak energy falls exponentially to its half of the initial value in about 0.6 Gyr. A curve is best fitted to the peak energy evolution using equation $f(x) = A\ \exp(-x/t) + B$. The best fit parameters are $A = 7.73 \times\ 10^{15}$ and $B = 7.5 \times\ 10^{15}$ with χ^2/DoF value of 6.7×10^{28}. On the other hand radius of the blast wave roughly seems to increase linearly with the time of its evolution.

4. Discussion

The main finding of this study is that the cluster merger phenomena morphologically and energetically almost similar to the blast waves. This shock when propagated near to or beyond the virial radius of the cluster and interacted with the cluster surroundings (see Fig 1, it took different shape depending on whether it meets the void/ accretion shocks or the filamentary inflow. When it meets the filaments the shock fronts get destroyed and at the surface of interaction the medium becomes turbulent due to Kelvin-Helmhoz instability and looses the thermal energy and create broken parts on the thermal shock fronts. On the other hand the shock strength is enhanced when it interacted to the accretion shock at the meeting surface of inter-cluster medium and void. This region is thus become the most efficient particle accelerator and magnetic field is magnified due to high compression rate. As a result, these regions produce significant amount of synchrotron radio emission due to acceleration of charge particles in the enhanced magnetic field, and curved shocks are made visible in radio waves. A ringlike formation with several broken bow shocks are thus observed. This study thus tries to connect the cluster merger shock wave and the formation of curved and nearly symmetric radio emission found in Abell 3376, Abell 3667, CIZA J2242.8+5301 etc. clusters.

Approach used here is crude as the code doesn't include magnetic field and particle acceleration physics. Still the analysis performed on the basis of energy evolution, shock compression and turbulent kinetic energy, provides interesting results in the comparison between these simulations and observations of ring like formations. I believe that even with simplified, purely hydro-dynamical approach I am able to show a first comparison with observed structures.

Acknowledgments

Computations described in this work were performed using the Enzo code (http://enzo.googlecode.com), and analysis done with yt-tool (http://yt-project.org). IUCAA high performance computing facility is used as computational resources. I am thankful to Joydeep Bagchi and Prakash Sarkar of IUCAA for discussion on the subject and providing technical help.

References

[1] Joydeep Bagchi, Florence Durret, Gastao B. Lima Neto and Surajit Paul 2006 *Science* 314 791
[2] Blandford, R., & Eichler, D. 1987, *Physics Report* 154 1
[3] Subramanian, K., Shukurov, A., & Haugen, N. E. L. 2006, *MNRAS*, 366, 1437
[4] Brunetti, G., et al., 2007, *MNRAS* 378 245
[5] Vazza, F., Tormen, G., Cassano, R., Brunetti, G., & Dolag, K. 2006,*MNRAS* 369 L14
[6] S. Paul, L. Iapichino, F. Miniati, J. Bagchi & K. Mannheim 2011 *Astophysical Journal* 726 17
[7] OShea, B. W., Bryan, G., Bordner, J., Norman, M. L., Abel, T., Harkness, R., & Kritsuk, A. 2005a *in Plewa T., Linde T., Weirs V.G, eds, Lecture Notes in Computational Science and Engineering* Vol. 41 *Adaptive Mesh Refinement Theory and Applications. Springer, Berlin, New York* 341
[8] E. Komatsu et al. [WMAP Collaboration], 2009 *Astrophys. J. Suppl.* 180 330
[9] Eisenstein, D. J., & Hu, W. 1999 *Astophysical Journal* 511 5
[10] Sarazin, C., & White, III R.E., 1987 *Astophysical Journal* 320 32
[11] Roettiger, K., Burns, J., & Loken, C. 1993 *Astophysical Journal* 407 L53

International Conference on Modern Perspectives of Cosmology and Gravitation (COSGRAV12) IOP Publishing
Journal of Physics: Conference Series **405** (2012) 012027 doi:10.1088/1742-6596/405/1/012027

ISCO and Principal Null Congruences in Extremal Kerr Spacetime

Parthapratim Pradhan
Department of Physics, Vivekananda Satabarshiki Mahavidyalaya, Manikpara, Paschim Medinipur, West Bengal 721513, India
E-mail: pppradhan77@gmail.com

Abstract. The effective potential in universal like coordinates(U, V, θ, ϕ), which are smooth across the event horizon is derived and investigated the ISCO(Innermost Stable Circular Orbits) explicitly in these coordinates for extremal Kerr spacetime. Extremization of the effective potential for timelike circular orbit shows that the existence of a stable circular geodesics in the extremal spacetime for direct orbit, precisely *on* the event horizon in terms of the radial coordinate which coincides with the *principal null geodesic congruences* of the event horizon. These null geodesic congruences mold themselves to the spacetime curvature in such a way that Weyl conformal tensor and its dual are vanished, that is why they are in-fact *doubly degenerate principal null congruences*.

1. Introduction:

Generic non extremal black hole spacetime is a different class than the extremal spacetime, it has been proved by several authors for Reissner Nordström(RN) blackhole[1, 3] and Kerr blackhole[2]. There are certain number of features which has been present in the non extremal case but in the extremal case they are completely absent. For example the 'bifurcate 2-sphere' which exists in near extremal spacetime to compare the 'macroscopic' entropy of a class of extremal black holes solutions of the supergravity limit of string theories, to the 'microscopic' entropy obtained from counting of string states[5] but the extremal spacetimes does not have any such 'bifurcate 2-sphere'.

It is well known that extremal spacetime do not have any trapped surface[1, 2] inside the event horizon (itself usually a marginal outer trapped surface), whereas the near extremal black hole spacetime have filled up with the trapped surface. Indeed, it is also true that[6] the proper spatial distance on a constant time slices from event horizon to Cauchy horizon in the extremal Kerr geometry is actually infinity but in the near extremal situation the proper distance is finite. It has been found that[2] in the extremal geometry, there exists a class of stable circular orbit so called ISCO on the event horizon of the extremal spacetime which coincides with the null geodesic generator of the event horizon[1, 2] but the near extremal black hole spacetime does not admit this class of geodesics. Lastly, the extremal black hole spacetime have zero Hawking temperature due to the zero surface gravity which measures the equilibrium temperature for the thermal distribution of the radiation, whereas the non extremal blackhole possesses Hawking temperature.

In this *letter*, we show an interesting feature of the *Kerr* spacetime, is that the effective potential for timelike circular geodesics and null circular geodesics *coalesce into zero energy*

Published under licence by IOP Publishing Ltd

geodesics and therefore direct ISCO coalesce with the *principal null geodesic congruences*. This happens only in extremal spacetime, near extremal spacetime does not admit such type of features. In particular, the class of geodesics close to the event horizon in the extremal case does not overlap with the class in the case infinitesimally close to the extremal.

The plan of the paper is as follows: In section 2 we explicitly derive the effective potential in universal like coordinates and discuss the subtleties of the coordinates on the horizon. In section 3 we calculate the principal null geodesic congruences . Finally, we conclude our discussions in section 5.

Consider the Kerr metric in Boyer-Lindquist coordinates,

$$ds^2 = -\frac{\Delta}{\rho^2}\left[dt - a\sin^2\theta d\phi\right]^2 + \frac{\sin^2\theta}{\rho^2}\left[(r^2+a^2)\,d\phi - adt\right]^2 + \rho^2\left[\frac{dr^2}{\Delta} + d\theta^2\right] . \quad (1)$$

where $a \equiv \frac{J}{M}$, $\rho^2 \equiv r^2 + a^2\cos^2\theta$ $\Delta \equiv r^2 - 2Mr + a^2 \equiv (r-r_+)(r-r_-)$. The horizon occurs at $r_\pm = M \pm \sqrt{M^2 - a^2}$. The outer horizon r_+ is called event horizon and the inner horizon r_- is called Cauchy horizon.

2. Zero Energy Geodesics:

The extremal Kerr metric is adapted from the work [2](See for details: arXiv:1108.2333v1). Therefore on the equatorial plane $\theta = \pi/2$ the metric can be written as

$$ds^2 = \mathcal{A}\left(\sec^4 U\,dU^2 + \csc^4 V\,dV^2\right) + \mathcal{B}\sec^2 U\,\csc^2 V dU dV + \mathcal{C}\,(d\phi^\star)^2$$
$$+\mathcal{D}\left(\sec^2 U\,dU + \csc^2 V\,dV\right)d\phi^\star . \quad (2)$$

where

$$\mathcal{A} = \frac{1}{8}(1-\frac{M}{r})^2\left(\frac{r^2}{r^2+M^2}+\frac{1}{2}\right)\left(\frac{r^2-M^2}{r^2+M^2}\right) + \frac{(M^2-r^2)^2}{16M^2r^2} . \quad (3)$$

$$\mathcal{B} = \frac{(M^2-r^2)^2}{8M^2r^2} - \frac{1}{2}(1-\frac{M}{r})^2\left[\frac{r^4}{(r^2+M^2)^2}+\frac{1}{4}\right] . \quad (4)$$

$$\mathcal{C} = \frac{(r^2+M^2)^2}{r^2} - M^2(1-\frac{M}{r})^2 , \mathcal{D} = \frac{1}{2}M(1-\frac{M}{r})^2 - \frac{(M^4-r^4)}{2Mr^2} . \quad (5)$$

The spacetime (2) has a timelike isometry generated by the timelike Killing vector field ξ whose projection along the 4-velocity $\mathbf{u}$ of geodesics: $\xi\cdot\mathbf{u} = -\mathcal{E}$, is conserved along such geodesics. Now, ξ has non-vanishing components ξ^U, ξ^V which can be easily derived from the fact that in the Schwarzschild coordinate basis $\xi = \partial_t$. One obtains $\xi^U = \cos^2 U$, $\xi^V = \sin^2 V$. Thus, in this coordinate chart, $\mathcal{E}$ can be expressed as $\mathcal{E} = -(\mathcal{A}+\mathcal{B}/2)\left[\sec^2 U\,u^U + \csc^2 V\,u^V\right] + 2\mathcal{D}\,u^\phi$. There is also the 'angular momentum' $L \equiv \zeta\cdot\mathbf{u}$ (where $\zeta \equiv \partial_\phi$) which is similarly conserved. It can be expressed as $L = \mathcal{D}\left[\sec^2 U\,u^U + \csc^2 V\,u^V\right] + \mathcal{C}\,u^\phi$. From normalization condition

$$u^2 = \mathcal{A}\left[\sec^4 U\,(u^U)^2 + \csc^4 V\,(u^V)^2\right] + \mathcal{B}\,\sec^2 U\,\csc^2 V\,u^U\,u^V$$
$$+\mathcal{D}\left[\sec^2 U\,u^U + \csc^2 V\,u^V\right]u^\phi + \mathcal{C}\,(u^\phi)^2 . \quad (6)$$

Journal of Physics: Conference Series **405** (2012) 012027 doi:10.1088/1742-6596/405/1/012027

For *circular* geodesics; the radial component of the 4-velocity vanishes: $u^r = 0$. In terms of components in our chosen coordinate chart, this translates into $r_U u^U + r_V u^V = 0$ where $r_U \equiv \partial r/\partial U$ etc. So that one can obtain for circular geodesics

$$\frac{u^U}{u^V} = \cos^2 U \csc^2 V \, . \tag{7}$$

$$\mathcal{E} = -\left[2(\mathcal{A}+\mathcal{B}/2)\sec^2 U\, u^U + 2\mathcal{D}\, u^\phi\right] . \tag{8}$$

$$L = 2\mathcal{D}\sec^2 U\, u^U + \mathcal{C}\, u^\phi \, . \tag{9}$$

From normalization of the geodesics equation we get

$$\mathbf{u}^2 = -(u^U \sec^2 U)\,\mathcal{E} + \mathcal{C}\,(u^\phi)^2. \tag{10}$$

For timelike circular geodesics $\mathbf{u}^2 = -1$, Using (8, 9) one can obtain the energy equation for timelike circular geodesics is

$$\alpha\,\mathcal{E}^2 + \beta\,\mathcal{E} + \gamma = 0\, . \tag{11}$$

where

$$\alpha = G\,\mathcal{C}\,\mathcal{D}^2 - \mathcal{C}\,\mathcal{H}^2, \quad \beta = 2\mathcal{C}\,\mathcal{D}GL\,(\mathcal{A}+\mathcal{B}/2) - 2\mathcal{D}L\mathcal{H}^2$$
$$\gamma = \mathcal{C}\,GL^2\,(\mathcal{A}+\mathcal{B}/2)^2 - G\,\mathcal{H}^2, \quad G = 4\mathcal{D}^2 - 2\mathcal{C}\,(\mathcal{A}+\mathcal{B}/2), \quad \mathcal{H} = \mathcal{C}\,(\mathcal{A}+\mathcal{B}/2) - 4\mathcal{D}^2\, . \tag{12}$$

Therefore the effective potential for timelike circular geodesics may be written as

$$\mathcal{E} = (\mathcal{V}_{eff})_{Horizon} = \frac{-\beta + \sqrt{\beta^2 - 4\alpha\gamma}}{2\alpha}\, . \tag{13}$$

Similarly the effective potential for null circular geodesics can be written as

$$\mathcal{E} = (\mathcal{U}_{eff})_{Horizon} = \frac{-\beta + \sqrt{\beta^2 - 4\alpha\gamma_0}}{2\alpha}\, . \tag{14}$$

where $\gamma_0 = \mathcal{C}\,GL^2\,(\mathcal{A}+\mathcal{B}/2)^2$

Observe that the future horizon of the spacetime is given by $U = \pi/2$ with V arbitrary: in other words $r(\pi/2, V) = M$, which is the direct ISCO for extremal Kerr blackhole also. One can compute the derivatives, it turns out that $r_U(\pi/2, V) = 2M^2$, while the other derivative of r is regular on the horizon. Now on the future horizon or at $r = M$(direct ISCO)

$$\mathcal{A}\to 0,\ \ \mathcal{B}\to 0,\ \ \mathcal{C}\to 4M^2,\ \ \mathcal{D}\to 0,\ \ \alpha\to 0,\ \ \beta\to 0,\ \ \gamma\to 0,\ \ G\to 0,\ \ \mathcal{H}\to 0$$
$$\mathcal{E}\to 0,\ \ \frac{u^U}{u^V} = \cos^2 U\csc^2 V \to 0\, . \tag{15}$$

which implies $u^U \to 0$ and $u^V \to \infty$ on the horizon. It follows that L is a finite quantity which indeed does not vanish on the horizon. It can also easily seen that the effective potential for timelike circular geodesics and null circular geodesics *coalesce into zero energy geodesics.* i.e.

$$\mathcal{E} = (\mathcal{V}_{eff})_{Horizon} = (\mathcal{U}_{eff})_{Horizon} \to 0\, . \tag{16}$$

Since both massive particles and photons are coincident to the zero energy geodesics, therefore they must mold themselves to the spacetime curvature. Thus Weyl conformal tensor and its dual are vanished on that curvature, so they are called doubly degenerate principal null congruences.

Journal of Physics: Conference Series **405** (2012) 012027 doi:10.1088/1742-6596/405/1/012027

3. Principal Null Congruences:

In Boyer-Lindquist coordinates the principal null congruences are

$$\begin{aligned} k^t &\equiv \frac{dt}{d\lambda} = \frac{(r^2+a^2)E}{\Delta} \\ k^r &\equiv \frac{dr}{d\lambda} = \pm E \\ k^\theta &\equiv \frac{d\theta}{d\lambda} = 0 \\ k^\phi &\equiv \frac{d\phi}{d\lambda} = \frac{aE}{\Delta} \ . \end{aligned} \tag{17}$$

These photon geodesics mold themselves to the spacetime curvature in such a way that, if $C_{\alpha\beta\gamma\delta}$ is the Weyl conformal tensor and $*C_{\alpha\beta\gamma\delta} = \epsilon_{\alpha\beta\mu\nu}C^{\mu\nu}_{\gamma\delta}$ is its dual, then

$$C_{\alpha\beta\gamma}[\delta k_\epsilon]k^\beta k^\nu = 0\,, *C_{\alpha\beta\gamma}[\delta k_\epsilon]k^\beta k^\nu = 0\ . \tag{18}$$

This implies, these photon geodesics are doubly degenerate principal null congruences.

The fact that the proper spatial distance[2] from ISCO to horizon tends to non zero and diverges at the extremal limit. But the proper time interval for in-falling radial geodesics for zero angular momentum on the horizon at the extremal limit will be $\frac{dr}{d\tau} = \pm\sqrt{\frac{2M}{r}(1+\frac{a^2}{r^2})}$. Now at the horizon at the extremal limit this can be $\frac{dr}{d\tau} = \pm 2$. $+$ sign for outgoing geodesics and $-$ sign for ingoing geodesics. Therefore in-falling proper time from photon orbit or ISCO or marginal bound orbit to the horizon becomes zero. This implies the fact that these orbits both coincides to the principal null congruences of the future event horizon.

4. Discussion:

Therefore the issue of direct ISCO on the horizon must *coincides* with the principal null geodesic congruences of the event horizon. Thus the effective potential for timelike circular geodesics and null circular geodesics *coalesce into zero energy geodesics* . This zero energy geodesics are hovering on the horizon but does not fall in or out. They eventually mold themselves to the spacetime curvature. This happens only in extremal blackhole spacetime. The possible existence of this zero energy trajectory on the event horizon of an extremal Kerr spacetime is a surprising feature. The evidence of this is that it appears as a global minimum of the effective potential. Notice that, the location of the ISCO on the event horizon is clearly indicated by the effective potential, its actual existence may illusory[7], since generically the black hole event horizon has a spatial foliation as a marginal outer trapped surface. Further, it is the limit of an infinite set of outer trapped surface 'inside' it i.e. for $r < M$. This implies that any stable circular orbit of massive particle and null circular orbit of massless particles on and encircling the horizon must end up with the null geodesic generators of the future horizon.

[1] P. Pradhan, P. Majumdar, *Physics Letters* **A 375** (2011) 474-479.
[2] P. Pradhan, P. Majumdar, arXiv:1108.2333v1 [gr-qc].
[3] P. Pradhan, ArXiv:1205.5656[gr-qc].
[4] R. M. Wald, *Phys. Rev.* **D 48** (1993) 3427.
[5] A. Sen, *Int. J. Mod. Phys.* **A24** (2009) 4225.
[6] J. D. Bekenstein, *Phys. Rev.* **D 7** (1973) 2333.
[7] J. M. Bardeen, W. H. Press, S. A. Teukolsky *The Astrophysical Journal* **178** (1972) 347-369.
[8] B. Carter, *Phys. Review. Lett.* **141**(4) (1966) 1242.
[9] S. Chandrashekar, *The Mathematical Theory of Black Holes*, Clarendon Press, Oxford (1983).

Trivial symmetries in a 3D topological torsion model of gravity

Rabin Banerjee[a,1] **and Debraj Roy**[a,2]
[a]SN Bose National Centre for Basic Sciences, Salt Lake, Blk JD, Sec 3, Kolkata - 700098.
E-mail: [1]rabin@bose.res.in, [2]debraj@bose.res.in

Abstract. We study the gauge symmetries in a Mielke-Baekler type model of gravity in 2+1 dimensions. The model is built in a Poincare gauge theory framework where localisation of Poincare symmetries lead to gravity. However, explicit construction of gauge symmetries in the model through a Hamiltonian procedure yields an apparently different set of symmetries, as has been noted by various authors. Here, we show that the two sets of symmetries are actually equivalent in a canonical sense, their difference being just a set of trivial symmetries.

1. Introduction

Gauge symmetries and their systematic construction, while being of great significance, is fraught with subtleties. They must be off-shell, i.e. the action should remain invariant under them without use of any equations of motion and there are systematic algorithms for construction of such off-shell symmetries [1, 2, 3]. In models of gravity obtained by gauging Poincare symmetries, it is however seen that such systematic approaches yield the Poincare symmetries only on-shell. So, at the off-shell level, it appears that there are two independent set of symmetries. This is unsettling. We would then be required to find additional independent first-class constraints to support these symmetries. However, physically the only independent symmetries present are the space-time translations and Lorentz boosts, i.e. the Poincare symmetries.

We set up the model in the Poincare gauge theory (PGT) formalism [4, 5, 6]. The global Poincare symmetries on a set of global coordinates x^μ are localised at every spacetime point by introducing local frames having coordinates x^i. Both sets are 3-dimensional for our model, though the construction can be carried out in any dimensions. To change physical variables from one set of coordinates to another, triad fields are introduced, which are defined through

$$g_{\mu\nu} = b^i{}_\mu b^j{}_\nu \eta_{ij} .$$

Here 'g' and 'η' are the metrics of the global and local sets of coordinates, respectively and 'η' is the standard Minkowskian metric with signature $(+,-,-)$. Now, to keep the invariance of the theory, a gauge connection field '$\omega^i{}_\mu$' is introduced through a newly defined gauge covariant derivative. For an arbitrary vector A^i, the derivative is written as

$$\nabla_\mu A^i = \partial_\mu A^i + \epsilon^i{}_{jk}\omega^j{}_\mu A^k .$$

The corresponding field strengths give rise to gravitational fields

$$[\nabla_\mu, \nabla_\nu]\, b^i{}_\rho = \epsilon^i{}_{jk} R^j{}_{\mu\nu} b^k{}_\rho ,$$

Published under licence by IOP Publishing Ltd

Journal of Physics: Conference Series **405** (2012) 012028 doi:10.1088/1742-6596/405/1/012028

where $R^i{}_{\mu\nu} = \partial_\mu\omega^i{}_\nu - \partial_\mu\omega^i{}_\nu + \epsilon^i{}_{jk}\,\omega^j{}_\mu\omega^k{}_\nu$ defines the Riemann. Similarly, the commutator of ∇_i derivatives give rise to the torsion $T^i{}_{\mu\nu} = \nabla_\mu b^i{}_\nu - \nabla_\nu b^i{}_\mu$. It must be emphasised that this is a setting in the Riemann-Cartan spacetime rather than the Riemann spacetime of usual general relativity. In general it contains torsion, which goes to zero only after imposition of the connection variable's equation of motion, giving back the torsion free Riemannian manifold of general relativity. The Poincare symmetries of the basic fields are [7]

$$\begin{aligned}\delta_{PGT}b^i{}_\mu &= -\epsilon^i{}_{jk}b^j{}_\mu\theta^k - \partial_\mu\xi^\rho\, b^i{}_\rho - \xi^\rho\,\partial_\rho b^i{}_\mu \\ \delta_{PGT}\omega^i{}_\mu &= -\partial_\mu\theta^i - \epsilon^i{}_{jk}\omega^j{}_\mu\theta^k - \partial_\mu\xi^\rho\,\omega^i{}_\rho - \xi^\rho\,\partial_\rho\omega^i{}_\mu\,,\end{aligned} \tag{1}$$

where θ^i parametrizes local Lorentz rotations and ξ^ρ local coordinate translations. They together constitute the $3+3=6$ independent Poincare symmetries in 3D. These symmetries are independent of the particular action being considered, so long as they are constructed in a proper tensorial manner.

Now, symmetries of an action can be constructed via a Hamiltonian analysis through completely off-shell procedures, as stated before. After this is carried out, we find the symmetries which keeps the action invariant, as expected. However they cannot be identified with the Poincare symmetries (1), even after an appropriate field-dependent redefinition of the symmetry parameters. The clue of identification, as we show, lies in 'trivial' gauge symmetries [8, 9]. They are symmetries of the form

$$\delta q_i = \Lambda_{ij}\,\frac{\delta S}{\delta q_j} \qquad (\Lambda_{ij} = -\Lambda_{ji})\,, \tag{2}$$

which keep any action *off-shell* invariant due to antisymmetry in the coefficients Λ_{ij}. Thus

$$\delta S = \frac{\delta S}{\delta q_i}\delta q_i = \frac{\delta S}{\delta q_i}\,\Lambda_{ij}\,\frac{\delta S}{\delta q_j} = 0 \tag{3}$$

without depending on the form of the Euler derivatives $\frac{\delta S}{\delta q_i}$ and without requiring the imposition of the equations of motion $\frac{\delta S}{\delta q_i} = 0$.

In what follows, we take up the Mielke-Baekler model, summarise its Hamiltonian constraints, outline construction of an off-shell gauge generator[1] and generate gauge symmetries. These are then compared with the Poincare symmetries to see how the two may actually be equivalent.

2. Hamiltonain analysis and construction of a gauge generator

The Mielke-Baekler 3D topological gravity model with torsion [11] along with a cosmological term is described by

$$S = \int \mathrm{d}^3\mathrm{x}\,\epsilon^{\mu\nu\rho}\Big[ab^i{}_\mu R_{i\nu\rho} - \frac{\Lambda}{3}\epsilon_{ijk}b^i{}_\mu b^j{}_\nu b^k{}_\rho + \alpha_3\Big(\omega^i{}_\mu\partial_\nu\omega_{i\rho} + \frac{1}{3}\epsilon_{ijk}\,\omega^i{}_\mu\omega^j{}_\nu\omega^k{}_\rho\Big) + \frac{\alpha_4}{2}b^i{}_\mu T_{i\nu\rho}\Big] \tag{4}$$

whose equations of motion are found by setting the Euler derivatives to zero

$$\begin{aligned}\frac{\delta S}{\delta b^i{}_\mu} &= \epsilon^{\mu\nu\rho}\left[a\,R_{i\nu\rho} + \alpha_4\,T_{i\nu\rho} - \Lambda\,\epsilon_{ijk}b^j{}_\nu b^k{}_\rho\right] = 0 \\ \frac{\delta S}{\delta\omega^i{}_\mu} &= \epsilon^{\mu\nu\rho}\left[\alpha_3\,R_{i\nu\rho} + a\,T_{i\nu\rho} + \alpha_4\,\epsilon_{ijk}b^j{}_\nu b^k{}_\rho\right] = 0.\end{aligned} \tag{5}$$

The constraints in the above model are [12] listed in Table 1 with the definitions

[1] Note that there exist methods [10] which construct symmetries of fields that keep the equations of motion invariant. We are using a different approach.

Table 1. Constraints of the theory.

	First Class ψ	Second class χ
Primary	$\phi_i{}^0$, $\Phi_i{}^0$	$\phi_i{}^\alpha$, $\Phi_i{}^\alpha$
Secondary	$\bar{\mathcal{H}}_i$, $\bar{\mathcal{K}}_i$	

$$
\begin{aligned}
\phi_i{}^\mu &= \pi_i{}^\mu - \alpha_4\,\epsilon^{0\alpha\beta}\,b_{i\beta}\,\delta^\mu_\alpha \\
\Phi_i{}^\mu &= \Pi_i{}^\mu - \epsilon^{0\alpha\beta}\,(2a\,b_{i\beta} + \alpha_3\,\omega_{i\beta})\,\delta^\mu_\alpha \\
\bar{\mathcal{H}}_i &= -\left[\epsilon^{0\alpha\beta}\left(a\,R_{i\alpha\beta} + \alpha_4\,T_{i\alpha\beta} - \Lambda\epsilon_{ijk}b^j{}_\alpha b^k{}_\beta\right)\right] - \nabla_\alpha\phi_i{}^\alpha + \epsilon_{ijk}\,b^j{}_\alpha\left(p\,\phi^{k\alpha} + q\,\Phi^{k\alpha}\right) \\
\bar{\mathcal{K}}_i &= -\left[\epsilon^{0\alpha\beta}\left(a\,T_{i\alpha\beta} + \alpha_3\,R_{i\alpha\beta} + \alpha_4\,\epsilon_{ijk}b^j{}_\alpha b^k{}_\beta\right)\right] - \nabla_\alpha\Phi_i{}^\alpha - \epsilon_{ijk}\,b^j{}_\alpha\phi^{k\alpha} \\
p &= \frac{\alpha_3\Lambda + \alpha_4 a}{\alpha_3\alpha_4 - a^2}\,; \qquad q = -\frac{\alpha_4^2 + a\Lambda}{\alpha_3\alpha_4 - a^2}.
\end{aligned} \tag{6}
$$

The gauge generator G which generates gauge symmetries is, by the Dirac principle, a linear combination of all first class constraints

$$G = \varepsilon_A\psi_A\,,$$

the sum in A running over the complete first-class sector. The symmetries of the basic variables are then obtained through $\delta q = \{q, G\}^* = \varepsilon_A\{q, \psi_A\}^*$ where $\{\,,\,\}^*$ indicates Dirac brackets obtained through elimination of the second class sector χ [7].

Now, we know that all the gauge parameters ε_A are not independent, as the number of independent gauge parameters must equal the number of independent primary first-class constraints. In the off-shell method that we adopt [1, 2, 3], the restriction on ε_A is obtained from the commutativity of gauge variations and time evolution $\delta\bullet\frac{d}{dt} \equiv \frac{d}{dt}\bullet\delta$. Explicit calculations [7] yield the generator

$$
\begin{aligned}
G = \int \mathrm{d}^2x\,\Big(\dot{\tau}^i\,\pi_i{}^0 + \tau^i&\left[\bar{\mathcal{H}}_i - \varepsilon_{ijk}(\omega^j{}_0 - p\,b^j{}_0)\pi^{k0} + q\,\varepsilon_{ijk}\,b^j{}_0\Pi^{k0}\right] \\
&+\dot{\sigma}^i\Pi_i{}^0 + \sigma^i\left[\bar{\mathcal{K}}_i - \varepsilon_{ijk}(b^j{}_0\,\pi^{k0} + \omega^j{}_0\,\Pi^{k0})\right]\Big)
\end{aligned} \tag{7}
$$

where τ^i and σ^i are six independent gauge parameters. We are now ready to study the gauge symmetries of the model utilising this generator.

3. Hamiltonian and Poincare gauge: role of trivial symmetries

The generator (7) yields the following symmetries of the basic fields

$$
\begin{aligned}
\delta_H b^i{}_\mu &= \nabla_\mu\tau^i - p\,\epsilon^i{}_{jk}\,b^j{}_\mu\tau^k + \epsilon^i{}_{jk}\,b^j{}_\mu\sigma^k, \\
\delta_H \omega^i{}_\mu &= \nabla_\mu\sigma^i - q\,\epsilon^i{}_{jk}\,b^j{}_\mu\tau^k.
\end{aligned} \tag{8}
$$

A first comparison of the two sets of symmetries (1) and (8) reveals that the Hamiltonian symmetries are explicitly dependant on the coupling parameters of the various terms in the action (4), in contrast to the PGT symmetries. Also, we note that the gauge parameters of the two symmetries are different. So we need to map the gauge parameters of (8) into the PGT gauge parameters. This is achieved through the field dependent map[2] [12]

$$\tau^i = -\xi^\rho b^i{}_\rho \qquad \& \qquad \sigma^i = -\theta^i - \xi^\rho \omega^i{}_\rho . \tag{9}$$

Using this map, the Hamiltonian symmetries can be written in the form

$$\delta_H \sim \delta_{PGT} + \text{equations of motion}.$$

It may seem that the two symmetries are thus equivalent only on-shell. To show that this is not the case, we write out their form explicitly

$$\begin{aligned} \delta_H b^i{}_\mu &= \delta_{PGT} b^i{}_\mu + \frac{\alpha_3}{2(\alpha_3\alpha_4 - a^2)} \eta^{ij} \xi^\rho \epsilon_{\mu\nu\rho} \frac{\delta S}{\delta b^j{}_\nu} - \frac{a}{2(\alpha_3\alpha_4 - a^2)} \eta^{ij} \xi^\rho \epsilon_{\mu\nu\rho} \frac{\delta S}{\delta \omega^j{}_\nu} \\ \delta_H \omega^i{}_\mu &= \delta_{PGT} \omega^i{}_\mu - \frac{a}{2(\alpha_3\alpha_4 - a^2)} \eta^{ij} \xi^\rho \epsilon_{\mu\nu\rho} \frac{\delta S}{\delta b^j{}_\nu} + \frac{\alpha_4}{2(\alpha_3\alpha_4 - a^2)} \eta^{ij} \xi^\rho \epsilon_{\mu\nu\rho} \frac{\delta S}{\delta \omega^j{}_\nu} \end{aligned} \tag{10}$$

To cast the balance terms of $\delta \equiv \delta_H - \delta_{PGT}$ into the form (2), we first write down the appropriate Λ matrix:

$$\begin{aligned} \Lambda_{(b^i{}_\mu, b^j{}_\nu)} &= \frac{\alpha_3}{2(\alpha_3\alpha_4 - a^2)} \eta^{ij} \xi^\rho \epsilon_{\mu\nu\rho} & \Lambda_{(b^i{}_\mu, \omega^j{}_\nu)} &= \frac{-a}{2(\alpha_3\alpha_4 - a^2)} \eta^{ij} \xi^\rho \epsilon_{\mu\nu\rho} \\ \Lambda_{(\omega^i{}_\mu, b^j{}_\nu)} &= \frac{-a}{2(\alpha_3\alpha_4 - a^2)} \eta^{ij} \xi^\rho \epsilon_{\mu\nu\rho} & \Lambda_{(\omega^i{}_\mu, \omega^j{}_\nu)} &= \frac{\alpha_4}{2(\alpha_3\alpha_4 - a^2)} \eta^{ij} \xi^\rho \epsilon_{\mu\nu\rho} . \end{aligned} \tag{11}$$

The anti-symmetry of the above structure can now be easily demonstrated. As an example,

$$\begin{aligned} \Lambda_{(b^i{}_\mu, \omega^j{}_\nu)} &= \frac{-a}{2(\alpha_3\alpha_4 - a^2)} \eta^{ij} \xi^\rho \epsilon_{\mu\nu\rho} \\ &= -\frac{-a}{2(\alpha_3\alpha_4 - a^2)} \eta^{ji} \xi^\rho \epsilon_{\nu\mu\rho} = -\Lambda_{(\omega^j{}_\nu, b^i{}_\mu)} . \end{aligned}$$

So, we see that indeed 'δ' is of the required form (2) for trivial gauge symmetries. Thus the invariance of the action can now be achieved under these 'balance' symmetries $\delta \equiv \delta_H - \delta_{PGT}$ without the use of any equations of motion. Hence we see, that from a canonical point of view,

$$\delta_H \equiv \delta_{PGT}$$

as their difference is just an un-physical 'trivial' symmetry.

References

[1] Henneaux M, Teitelboim C and Zanelli J 1990 Nucl. Phys. B **332**, 169-188.
[2] Banerjee R, Rothe H J and Rothe K D 1999 Phys. Lett. B **463** 248.
[3] Banerjee R, Rothe H J and Rothe K D 2000 Phys. Lett. B **479**, 429.
[4] Utiyama R 1956 Phys. Rev. **101** 1597.
[5] Kibble T W B 1961 J. Math. Phys. **2** 212.
[6] Sciama D W 1962 *Recent Developments in General Relativity, Festschrift for Leopold Infeld* (New York: Pergamon Press) p 415.
[7] Banerjee R, Gangopadhyay S, Mukherjee P and Roy D 2010 *J. High Energy Phys.* JHEP **1002**, 075 (2010)
[8] Henneaux M and Teitelboim C 1992 *Quantization of gauge systems* (Princeton, USA: Univ. Pr.)
[9] Henneaux M, Kleinschmidt A and Gomez G L 2010 Phys. Rev. D **81**, 064002.
[10] Castellani L 1982 Annals Phys. **143**, 357.
[11] Baekler P, Mielke E W and Hehl F W 1992 Nuovo Cim. B **107** 91.
[12] Blagojevic M and Cvetkovic B 2006 *Trends in GR and QC* **2** ed Ch Benton (New York: Nova Science) 103.
[13] Banerjee R and Roy D 2011 Phys. Rev. D **84**, 124034.

[2] For a construction of the map, using Noether identities corresponding to the two sets of symmetries, see [13].

Journal of Physics: Conference Series **405** (2012) 012029 doi:10.1088/1742-6596/405/1/012029

Noncommutative quantum mechanics of a test particle under linearly polarized gravitational waves

Anirban Saha

Department of Physics, West Bengal State University, Barasat, India

E-mail: anirban@iucaa.ernet.in

Abstract. We consider the quantum dynamics of a test particle first with no potential and then with hermonic oscillator potential in noncommutative space under the influence of a passing linearized gravitational wave in the long wave-length and low-velocity limit. A prescription for quantizing the classical Hamiltonian for these systems is proposed. While both the solutions show noncommutative signatures, for the hermonic oscillator case a resonance scenario is found. We expect this noncommutative signature to show up as some noise source in the GW detection experiments since the recent phenomenological upper-bounds set on spatial noncommutative parameter implies a length-scale comparable to the length-variations due to the passage of gravitational waves, detectable in the present day GW detectors.

Gravitational waves (GWs) are tiny disturbances in spacetime. Their effect on matter is expected to be measured by the relative optical phase shift between the light paths in two perpendicular km-length arm cavities caused by the tiny displacement of two test mass mirrors (hung at the end of each cavity) induced by passing GWs. Since the cavity arm length $L \sim 1\text{km}$, δL will be $\sim 10^{-18}\text{m}$. It is, therefore, at the quantum mechanical level that experimental evidence for the GWs is likely to appear [1].

Interestingly, in recent developments of noncommutative (NC) quantum mechanics and NC quantum field theory, where the coordinates x^{μ} satisfy the NC algebra $[x^{\mu}, x^{\nu}] = i\theta^{\mu\nu}$ the upperbounds on various NC parameters appearing in the literature [2, 3, 4, 5, 6, 7, 8] are quite close to this length scale. A wide range of theories have been constructed in a NC framework including various gauge theories [9], gravity [10] and even encompassing certain possible phenomenological consequences [11, 12, 13, 14, 15]. The upperbound on the value of the coordinate commutator θ^{ij} found in [2] is $(10\text{TeV})^{-2}$ which corresponds to $4 \times 10^{-40}\text{m}^2$ for $\hbar = c = 1$. Whereas such upperbounds on time-space NC parameter θ^{0i} is $9.51 \times 10^{-18}\text{m}^2$. However, recent studies in NC quantum mechanics revealed that this bound could be as high as $(4\text{GeV})^{-2} - (30\text{MeV})^{-2}$ [6]. These upperbounds correspond to the length scale $\sim 10^{-20}\text{m} - 10^{-17}\text{m}$.

With the prospect of the direct detection of GWs of such tiny amplitude as $\sim 10^{-18}$ in the near future, a possibility of detecting the NC structure of spacetime would be in the GW detection experiments. Therefore, we would construct the NC quantum mechanics of a free test particle and hermonic oscillator, interacting with a linearly polarised GW. It has been demonstrated [10] that any NC correction in the gravity sector is second order in the NC parameter, therefore, in a first order theory the linearised GW remains unaltered and any NC correction will be through the particle sector only. We incorporate the NC effect by writing the NC Hamiltonian and then reexpressing it in terms of the commutative coordinates and their momenta by the well known Bopp-shift transformations [6, 8]. Then we quantize the system following [16].

Journal of Physics: Conference Series **405** (2012) 012029 doi:10.1088/1742-6596/405/1/012029

Choosing the TT gauge removes the gauge redundancies and the GW is characterised by the non-zero + and × polarised components $h_{11} = -h_{22}$, and $h_{12} = h_{21}$ respectively. The non-trivial components of the curvature tensor and the geodesic deviation equation[1] governing the response of a 2-dimensional harmonic oscillator with frequency ϖ to the passage of a GW, in the proper detector frame, become [17]

$$R^j{}_{0,k0} = -\frac{\partial \Gamma^j_{0k}}{\partial t} = -\frac{1}{2}\frac{\partial^2 h_{jk}}{\partial t^2}; \qquad m\frac{d^2x^j}{dt^2} = -mR^j{}_{0,k0}x^k + F(x). \tag{1}$$

Here x^j is the proper distance of the test particle from the origin, m is its mass and $F(x) = -m\varpi^2 x^j$ is the restoring force. The Lagrangian for the system, upto a total derivative term [16], is $\mathcal{L} = \frac{1}{2}m\dot{x}^2 - m\Gamma^j{}_{0k}\dot{x}_j x^k - \frac{1}{2}m\varpi^2 x^2$ and the canonical momentum is $p_j = m\dot{x}_j - m\Gamma^j_{0k}x^k$. The Hamitonian for the system becomes

$$H = \frac{1}{2m}\left(p_j + m\Gamma^j_{0k}x^k\right)^2 + \frac{1}{2}m\varpi^2 {x_j}^2 . \tag{2}$$

For GWs propagating along the z-axis, $\Gamma^j{}_{0k}$ has non-zero components only in the $x-y$ plane and the particle moves freely in the z-direction. So we focus on the planar motion of the particle. To "quantize" this system in the NC plane, we replace x^j and p_j in the above Hamiltonian by operators $\hat{x}^j$ and $\hat{p}_j$ satisfying the NC Heisenberg algebra $[\hat{x}_i, \hat{p}_j] = i\hbar\delta_{ij}$, $[\hat{x}_i, \hat{x}_j] = i\theta\epsilon_{ij}$, $[\hat{p}_i, \hat{p}_j] = 0$. It is well known that this can be mapped to the standard ($\theta = 0$) Heisenberg algebra spanned by X_i and P_j using [8, 6]

$$\hat{x}_i = X_i - \frac{1}{2\hbar}\theta\epsilon_{ij}P_j\,, \quad \hat{p}_i = P_i\,. \tag{3}$$

Using the traceless property of the GW and rewriting the NC version of Eq.(2) in terms of the operators X_i and P_j, we obtain

$$\hat{H} = \frac{{P_j}^2}{2m} + \Gamma^j_{0k}X_jP_k - \frac{\theta}{2\hbar}\epsilon_{jm}P_mP_k\Gamma^j_{0k} + \frac{1}{2}m\varpi^2{X_j}^2 - \frac{m\varpi^2}{2\hbar}\theta\epsilon_{jm}X^jP_m\,. \tag{4}$$

Since we are dealing with linearized gravity, a term quadratic in Γ has been neglected. Removing the last term in r.h.s. of eqn.s (1, 2)and the last two terms in eqn.(4) gives the free test particle case, where the only NC term is coupled with the GW through $\Gamma^j{}_{0k} = \frac{1}{2}\dot{h}_{jk}$ and both θ and h_{jk} being extremly small we expect that NC signature, though present for a free particle case, will be too tiny to detect. We therefore focus on the harmonic oscillator case in the sequel.

Defining raising and lowering operators[2]

$$X_j = \left(\frac{\hbar}{2m\varpi}\right)^{1/2}\left(a_j + a_j^\dagger\right); \qquad P_j = -i\left(\frac{\hbar m\varpi}{2}\right)^{1/2}\left(a_j - a_j^\dagger\right) \tag{5}$$

we write the Hamiltonian (4) for harmonic oscillator interacting with GW as

$$\hat{H} = \frac{\hbar\varpi}{4}\left(2a_j^\dagger a_j + 1 - a_j^2 - {a_j^\dagger}^2\right) - \frac{i\hbar}{4}\dot{h}_{jk}\left(a_ja_k - a_j^\dagger a_k^\dagger\right) + \frac{m\varpi\theta}{8}\epsilon_{jm}\dot{h}_{jk}\left(a_ma_k - a_ma_k^\dagger + C.C\right) \tag{6}$$

where C.C means complex conjugate. Working in the Heisenberg representation, the time evolution of $a_j(t)$ is given by

$$\frac{da_j(t)}{dt} = -i\varpi a^j + \frac{1}{2}\dot{h}_{jk}a_k^\dagger - \frac{m\varpi^2\theta}{2\hbar}\epsilon_{jk}a_k + \frac{im\varpi\theta}{8\hbar}\left(\epsilon_{lj}\dot{h}_{lk} + \epsilon_{lk}\dot{h}_{lj}\right)\left(a_k - a_k^\dagger\right) \tag{7}$$

[1] It describes the evolution of proper distance in TT-gauge frame for small velocities and $|x^j| \ll \frac{\lambda}{2\pi}$, the reduced wavelength of GW; the so-called *small-velocity and long wavelength limit*. Latin indices run from $1-3$.

[2] For free the particle ϖ is determined from the initial uncertainty in either the position or the momentum.

and that of $a_j^\dagger(t)$ is the C.C of Eq.(7). Noting that the raising and lowering operators must satisfy the commutation relations $[a_j(t), a_k(t)] = 0$, $[a_j(t), a_k^\dagger(t)] = \delta_{jk}$, we write them in terms of $a_j(0)$ and $a_j^\dagger(0)$, by the time-dependent Bogoluibov transformations

$$a_j(t) = u_{jk}a_k(0) + v_{jk}a_k^\dagger(0); \qquad a_j^\dagger(t) = a_k^\dagger(0)\bar{u}_{kj} + a_k(0)\bar{v}_{kj} \tag{8}$$

where the bar denotes the C.C and u_{jk} and v_{jk} are the generalized Bogoluibov coefficients with the boundary conditions $u_{jk}(0) = I$ and $v_{jk}(0) = 0$. Then, from eq.(7) and its C.C we get the equations of motions in terms of $\xi = u + v^\dagger$ and $\zeta = u - v^\dagger$:

$$\frac{d\zeta_{jk}}{dt} = -i\varpi\xi_{jk} - \frac{1}{2}\dot{h}_{jl}\zeta_{lk} - \frac{m\varpi^2\theta}{2\hbar}\epsilon_{jl}\zeta_{lk}, \quad \frac{d\xi_{jk}}{dt} = -i\varpi\zeta_{jk} + \frac{1}{2}\dot{h}_{jl}\xi_{lk} + \Theta_{jl}\zeta_{lk} - \frac{m\varpi^2\theta}{2\hbar}\epsilon_{jl}\xi_{lk}, \tag{9}$$

In the two-dimensional plane, the linearly polarized GW h_{jk} is most conveniently written in terms of the first and third Pauli spin matrices as $h_{jk}(t) = 2f(t)\left(\varepsilon_\times\sigma^1_{jk} + \varepsilon_+\sigma^3_{jk}\right) = 2f(t)\varepsilon_A\sigma^A_{jk}$. $2f(t)$ is the amplitude of the GW, $\varepsilon_\times(t)$ and $\varepsilon_+(t)$ are the two polarization states of the GW. Assuming the GW hits the particle at $t = 0$ so that $f(t) = 0$, for $t \leq 0$. To solve Eqs.(9) we write ζ_{jk} and ξ_{jk} as linear combinations of the Pauli spin matrices and identity matrix:

$$\zeta_{jk}(t) = AI_{jk} + B_1\sigma^1_{jk} + B_2\sigma^2_{jk} + B_3\sigma^3_{jk}, \quad \xi_{jk}(t) = CI_{jk} + D_1\sigma^1_{jk} + D_2\sigma^2_{jk} + D_3\sigma^3_{jk} \tag{10}$$

Substituting for h_{jk}, ζ_{jk} and ξ_{jk} in Eqs.(9) and comparing the coefficients of I and σ-matrices, we get a set of first order differential equations for $A, B_1, B_2, B_3, C, D_1, D_2, D_3$. Noting that $|f(t)| << 1$, these equations are solved iteratively about its $f(t) = 0$ solution to first order in the GW amplitude $f(t)$ and the NC parameter $\Lambda = \frac{m\varpi^2\theta}{2\hbar}$ and we get

$$\begin{aligned}
A(t) &= C(t) = e^{-i\varpi t}\cos(\Lambda t), \quad B_2(t) = D_2(t) = -ie^{-i\varpi t}\sin(\Lambda t) \\
B_1(t) &= -D_1(t) = -e^{-i\varpi t}\left[(\varepsilon_1\cos(\Lambda t) - \varepsilon_3\sin(\Lambda t))f(t) + 2i\varpi\varepsilon_1\int_0^t dt' e^{i\varpi(t-t')}\cos(\Lambda t')f(t')\right. \\
&\quad \left. -2i\varpi\varepsilon_3\int_0^t dt' e^{i\varpi(t-t')}\sin(\Lambda t')f(t')\right] + \varpi^2\left(\varepsilon_1\int_0^t g_1(t')dt' - \varepsilon_3\int_0^t g_2(t')dt'\right) \\
B_3(t) &= -D_3(t) = -e^{-i\varpi t}\left[(\varepsilon_3\cos(\Lambda t) + \varepsilon_1\sin(\Lambda t))f(t) + 2i\varpi\varepsilon_3\int_0^t dt' e^{i\varpi(t-t')}\cos(\Lambda t')f(t')\right. \\
&\quad \left. +2i\varpi\varepsilon_1\int_0^t dt' e^{i\varpi(t-t')}\sin(\Lambda t')f(t')\right] + \varpi^2\left(\varepsilon_3\int_0^t g_1(t')dt' + \varepsilon_1\int_0^t g_2(t')dt'\right)
\end{aligned} \tag{11}$$

where $g_1(t) = \int_0^t dt' e^{-i\varpi t'}\cos(\Lambda t')f(t')$, $g_2(t) = \int_0^t dt' e^{-i\varpi t'}\sin(\Lambda t')f(t')$. The system has now been essentially solved. Combining the expressions for $A, B_1, B_2, B_3, C, D_1, D_2, D_3$, we can write the solutions for ζ and ξ which in turn give u and v. Using Eq.(8), we can now combine u and v into $a_j(t)$ and $a_j^\dagger(t)$. From the initial position and momentum expectation values, i.e.$\langle\vec{r}_0\rangle = (X_1(0), X_2(0))$ and $\langle\vec{P}_0\rangle = (P_1(0), P_2(0))$, we get the raising and lowering operator $a_j(0)$ and $a_j^\dagger(0)$ at initial time $t = 0$. We then use them in Eq.(s) (8) to find $a_j(t)$ and $a_j^\dagger(t)$ at a general time t and these yield the time evolution of the expectation values of position coordinates $\langle X_1(t)\rangle$ and $\langle X_2(t)\rangle$ of the pendulum. The general expression of $\langle X_1(t)\rangle$ thus obtained is given by

$$\begin{aligned}
\langle X_1(t)\rangle &= [Re(A) + Re(D_3^\star)]X_1(0) + [Im(D_2) + Re(D_1^\star)]X_2(0) + [-Im(A) + Im(D_3^\star)]\frac{P_1(0)}{m\varpi} \\
&+ [Re(D_2) + Im(D_1^\star)]\frac{P_2(0)}{m\varpi}
\end{aligned} \tag{12}$$

where Re and Im denote the real and imaginary parts of the A, D_1, D_2 and D_3. Substituting their values from the equation (11) will give the expression for $\langle X_1(t)\rangle$ for a general GW amplitude $f(t)$. Assuming a monochromatic GW waveform $f(t) = f_0 e^{i\varpi' t}$, with frequency ϖ', in equation ((12) gives

$$\begin{aligned}
\langle X_1\rangle &= \frac{(\cos\varpi_- t + \cos\varpi_+ t)}{2}X_1(0) + \frac{(\sin\varpi_- t + \sin\varpi_+ t)}{2m\varpi}P_1(0) + \frac{(\cos\varpi_+ t - \cos\varpi_- t)}{2}X_2(0) \\
&+ \frac{(\sin\varpi_+ t - \sin\varpi_- t)}{2m\varpi}P_2(0) + \left(1 + \frac{\varpi^2\Delta\varpi t}{\Delta\varpi^2 - \Lambda^2}\right) f_0\left\{\varepsilon_3 X_1(0) + \varepsilon_1 X_2(0)\right\} - \left(\frac{\varpi^2\Lambda t}{\Delta\varpi^2 - \Lambda^2}\right) \\
&\times\ f_0\left\{\varepsilon_1 X_1(0) + \varepsilon_3 X_2(0)\right\} - (\varepsilon_1 + \varepsilon_3) f_0\left[\left(\frac{\varpi'_+}{\Delta\varpi_-}\right)^2 \sin\frac{\Delta\varpi_- t}{2}\left(X_1(0)\sin\frac{\Delta\varpi_- t}{2}\right.\right. \\
&- \left.\frac{P_1(0)}{m\varpi}\cos\frac{\Delta\varpi_- t}{2}\right) + \left(\frac{\varpi'_-}{\Delta\varpi_+}\right)^2 \sin\frac{\Delta\varpi_+ t}{2}\left.\left(X_2(0)\sin\frac{\Delta\varpi_+ t}{2} - \frac{P_2(0)}{m\varpi}\cos\frac{\Delta\varpi_+ t}{2}\right)\right] \\
&+ (\varepsilon_1 - \varepsilon_3) f_0\left[\left(\frac{\varpi'_-}{\Delta\varpi_+}\right)^2 \sin\frac{\Delta\varpi_+ t}{2}\left(X_1(0)\sin\frac{\Delta\varpi_+ t}{2} - \frac{P_1(0)}{m\varpi}\cos\frac{\Delta\varpi_+ t}{2}\right)\right. \\
&- \left.\left(\frac{\varpi'_+}{\Delta\varpi_-}\right)^2 \sin\frac{\Delta\varpi_- t}{2}\left(X_2(0)\sin\frac{\Delta\varpi_- t}{2} - \frac{P_2(0)}{m\varpi}\cos\frac{\Delta\varpi_- t}{2}\right)\right] \qquad (13)
\end{aligned}$$

where $\Delta\varpi = \varpi - \varpi'$, $\varpi_\pm = \varpi \pm \Lambda$, $\varpi'_\pm = \varpi' \pm \Lambda$ and $\Delta\varpi_\pm = \Delta\varpi \pm \Lambda$. This result implies that the presence of noncommutativity alters the response of the harmonic oscillator to a periodic GW from its commutative counterpart. When the frequency of the GW is very close to that of the harmonic oscillator ($\Delta\varpi \approx 0$), the oscillatory terms present in the solution representing the response of the system to the GW will oscillate with frequency Λ with a large amplitude. It may be possible to detect this effect. Now putting $\theta = 0$, i.e. $\Lambda = 0$ gives us the classical result of GW interacting with a harmonic oscillator in the low-velocity, long-wavelength limit. Interestingly, the presence of $\frac{1}{\hbar}$ factor in Λ, i.e. in the NC correction terms even after the computation of the expectation value indicates that the NC effect is inherently quantum mechanical in nature.

[1] C. M. Caves, K. S. Thorne, R. W. P. Drever, V. D. Sandberg, M. Zimmermann, *Rev. Mod. Phys.* **52** 341 .
[2] Carroll S. M. *et al.* 1980 *Phys. Rev. Lett.* **87** 141601.
[3] Bertolami O. *et al.* 2005 *Phys. Rev. D* **72** 025010.
[4] Banerjee R.*et al.* 2006 *Phys. Rev. D* **74** 045015.
[5] Saha A. 2007 *Eur. Phys. J. C* **51** 199.
[6] Stern A. 2008 *Phys. Rev. Lett.* **100** 061601.
[7] Mocioiu I. *et al.* 2000 Phys. Lett. B **489** 390.
[8] Chaichian M. *et al.* 2001 *Phys. Rev. Lett.* **86** 2716.
[9] See for example, Szabo R. J. 2003 *Phys. Rept.* **378** 207 and the references therein.
[10] Mukherjee P.and Saha A. 2006 *Phys. Rev. D***74** 027702 and the references therein.
[11] Sheikh-Jabbari M. M. 1999*Phys. Lett. B* **455** 129; *ibid.* 2000 *Phys. Rev. Lett.* **84** 5625 .
[12] Najafabadi M. M. 2006 *Phys. Rev. D* **74** 025021.
[13] Abel S. A. *et al.* 2006 *JHEP* **0601** 105.
[14] Melic B. *et al.* 2005 *Phys. Rev. D* **2** 057502.
[15] Carone C. D. 2006 *J. Phys. Conf. Ser.* **37** 96 and the references therein.
[16] Speliotopoulos A. D. 1995 *Phys. Rev. D* **51** 1701.
[17] Miggiore M. *Gravitational Wave, Vol I, Theory and Experiments* Oxford University Press, 2008.

International Conference on Modern Perspectives of Cosmology and Gravitation (COSGRAV12) IOP Publishing
Journal of Physics: Conference Series **405** (2012) 012030 doi:10.1088/1742-6596/405/1/012030

Exploring the Cosmic Web in the Sloan Digital Sky Survey Data Release Six using the Local Dimension

Prakash Sarkar[1*]

[1] IUCAA, Post Bag 4, Ganeshkhind, Pune University Campus, Pune 411 007, India.

E-mail: *sarkar@iucaa.ernet.in

Abstract.
It is possible to visualize the Cosmic Web as an interconnected network of one dimensional filaments, two dimensional sheets and three dimensional volume filling structures which we refer to as clusters. We have used the Local Dimension D, which takes value $D = 1, 2$, and 3 for filaments, sheets and clusters respectively, to analyse the Cosmic Web in a three dimensional volume limited galaxy sample from the Sloan Digital Sky Survey Data Release Six. The analysis was carried out separately using three different ranges of length-scales, 0.5 to 5 h^{-1}Mpc, 1 to 10 h^{-1}Mpc and 5 to 50 h^{-1}Mpc. We find that there is a progressive increase in the D values as we move to larger length-scales. At the smallest length-scales the galaxies predominantly reside in filaments and sheets. There is a shift from filaments to sheets and clusters at larger scales. Filaments are completely absent at the largest length-scales (5 to 50 h^{-1}Mpc). Considering the effect of the density environment on the Cosmic Web, we find that the filaments preferentially inhabit regions with a lower density environment as compared to sheets and clusters which prefer relatively higher density environments. A similar length-scale dependence and environment dependence was also found in a galaxy sample drawn from the Millennium Simulation which was analysed in exactly the same way as the actual data.

1. Introduction

Galaxy redshift surveys (like Two-Degree Field Galaxy Redshift Survey (2dFGRS), [1], Sloan Digital Sky Survey (SDSS), [2]) map the distribution of galaxies in the Universe and tell us directly how the present Universe looks like. These surveys clearly demonstrate the distribution of galaxies in a interconnected complex network of filaments, sheets and clusters, encircling the empty voids, called Cosmic Web. Quantifying the Cosmic Web and understanding it's origin is one of the most interesting and challenging issues in cosmology.

There are wide variety of statistical measures ([3]) to quantify the Cosmic Web but each quantifies one or atmost a few aspects of the Cosmic Web. We have used the Local Dimension ([4], [5]) as a means to analyse the patterns in the galaxy distribution in the Sloan Digital Sky Survey Data Release Six (SDSS DR6). We have also carried out a similar analysis on a galaxy catalogue from the semi-analytic model of galaxy formation implemented in the Millennium Simulation [6], and compared the results from the actual data with those from the simulation. The Local Dimension classifies the different structural elements along the Cosmic Web as filaments, sheets and clusters. It is well accepted that the Cosmic Web is a complex network whose morphology and connectivity will depend on the length-scale at which the analysis is carried out. [7] shows that the SDSS DR6 is consistent with homogeneity at length-scales

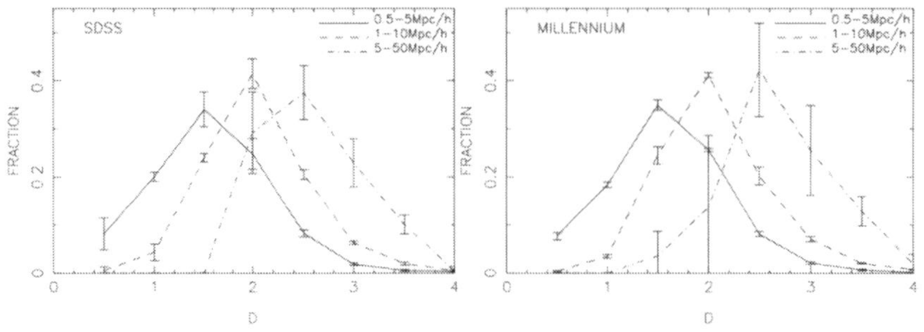

(a) (b)

Figure 1: This shows the fraction of galaxies with a particular D value for the SDSS data (left) and the Millennium data (right). The three different curves correspond to D values that were determined using the length-scales $0.5 - 5\,h^{-1}$ Mpc, $1 - 10\,h^{-1}$ Mpc and $5 - 50\,h^{-1}$ Mpc respectively. The bins in D have size ± 0.25.

beyond $60 - 70\,h^{-1}$ Mpc. We have carried out the entire analysis for three different ranges of length-scales namely 0.5 to 5 h^{-1}Mpc, 1 to 10 h^{-1}Mpc and 5 to 50 h^{-1}Mpc.

2. Data and Method of Analysis

2.1. Data

The analysis presented here is based on galaxy redshift data from the Main Galaxy Sample of SDSS DR6 [8]. The target selection algorithm of the Main Galaxy sample of SDSS DR6 is detailed in [9]. We have considered a volume-limited subsample in the Northern Galactic Cap which spans $50^\circ < \lambda < 30^\circ$ and $6^\circ < \eta < 35^\circ$, where λ and η are survey co-ordinates. The subsample contains 32740 galaxies in the $0.035 \le z \le 0.076$ (redshift range) and $20.5 \le M_r \le -19$ (absolute magnitude range). For estimating the cosmic variance of the data, we have generated ten bootstrap sample of the SDSS data.

We have also considered a semi analytic galaxy catalogue ([10]) from Millennium Run Simulation ([6]). The catalog contains around 9 million galaxies in a $(500\,h^{-1}\text{Mpc})^3$ box. Using the peculiar velocities, we map the galaxies to redshift space and then identify a region having the same geometry as our actual data. We have extracted three different sample from the Millennium Simulation.

2.2. Method of analysis

We choose a particular galaxy as center and determine the no of other galaxies $N(< R)$ inside a sphere of comoving radius R. This counting process is repeated by varying R. We assigned Poisson error $\Delta N(< R) = \sqrt{N(< R)}$ to each measured value of $N(< R)$. The Local Dimension D in the neighbourhood of the center is then determined by using a power-law fit of the equation $N(< R) \propto R^D$ over some length-scales $R_1 \le R \le R_2$, iff $\chi^2/\nu \le 1$. We have carried out the analysis for three different ranges of length-scales, each covering a decade. Values of D were determined separately across the length-scales 0.5 to 5 h^{-1}Mpc, 1 to 10 h^{-1}Mpc and 5 to 50 h^{-1}Mpc.

3. Results and Conclusions

We first analyse the fraction of centers with different D values, shown in Figure 1. The solid curves in the Figure 1a, corresponds to the results for 0.5 to 5 h^{-1}Mpc for the SDSS data, shows a broad peak with a maxima at $D = 1.5$. This bin together with bin centered at $D = 2$

Journal of Physics: Conference Series **405** (2012) 012030 doi:10.1088/1742-6596/405/1/012030

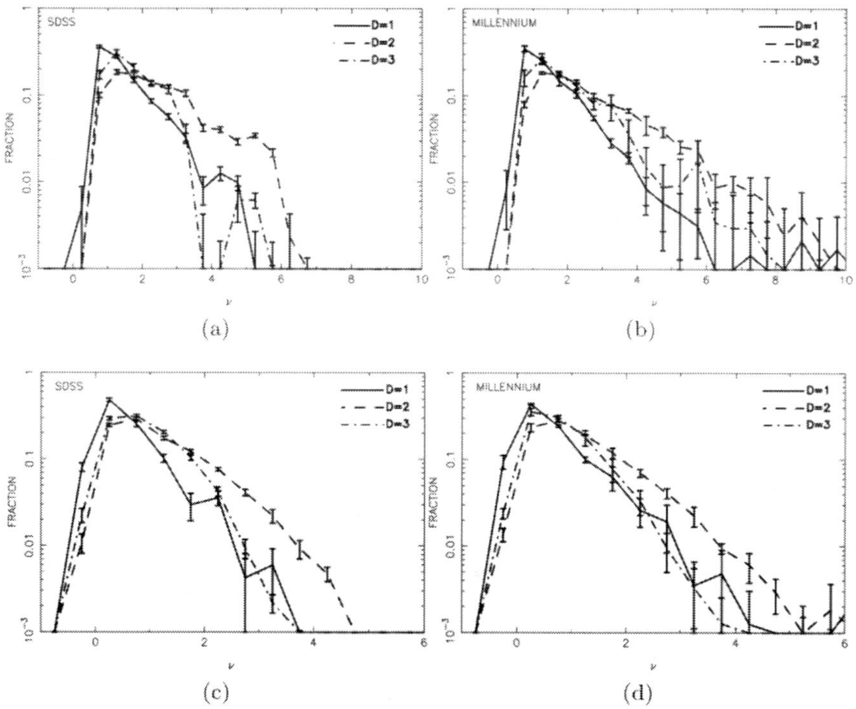

Figure 2: The three curves which correspond to $D = 1, 2$ and 3 respectively show the fraction of galaxies as a function of ν. Figure 2a and 2b are for the SDSS and Millennium data respectively using $R_1 = 0.5\,h^{-1}$ Mpc, $R_1 = 5\,h^{-1}$ Mpc. Figure 2c and 2d shows the same for $R_1 = 1\,h^{-1}$ Mpc, $R_1 = 10\,h^{-1}$ Mpc for SDSS and Millennium data respectively.

contains more than 50% of the centers. The dash curve in the Figure 1a, which corresponds to length-scales 1 to 10 h^{-1}Mpc, shows a sharp peak at $D = 2$. The three bins at $D = 1.5$, 2 and 2.5 together contains more than 80% of the centers. The dot-dashed curve in Figure 1, which corresponds to the length-scales 5 to 50 h^{-1}Mpc, shows a broader peak with maxima at $D = 2.5$. The two bins $D = 2$ and $D = 2.5$ contains more than 70% of the centers.

The above results implies that the majority of galaxies are predominantly contained in sheets and filaments at length-scales 0.5 to 5 h^{-1}Mpc, while at length-scale 1 to 10 h^{-1}Mpc Cosmic Web are predominantly in sheets. At length-scale 5 to 50 h^{-1}Mpc, the galaxies in Cosmic Web is predominantly in sheets and clusters and interestingly we don't find any centers with $D = 1$ or 1.5. The three curves shows that there is a shift to larger D values in Figure 1a as we progressively consider larger length-scales. It is quite evident from this that the nature of the structural elements that make up the Cosmic Web differs depending on the length-scale at which we view the Cosmic Web. For comparison, we have also applied the Local Dimension to analyse the galaxy distribution in the Millennium Simulation (Figure 1b). We find that the fraction of centers with different D values have very similar distributions in the actual SDSS data and the Millennium Simulation.

We now focus on the distribution of centers with a particular D values at different density environments. To determine the density environment, we consider the entire galaxy distribution to a density field on a grid of spacing $[0.5\,h^{-1}\text{Mpc}]^3$ using Cloud-in-Cell method. This density field is then smoothed with a Gaussian kernel having a smoothing length R_S. The density field at any grid point was quantified using the dimensionless ratio $\nu = \delta/\sigma$ where $\delta = \delta\rho/\bar{\rho}$

is the density contrast of the smoothed density field at the particular grid point and σ is the standard deviation of the density contrast of the smoothed density field evaluated using all the grid points that lie within the survey volume. The value of ν associated with any of the galaxies gives an estimate of the density environment in the vicinity of the structural element centered on that galaxy. Figure 2a shows the results for the range of length-scale $R_1 = 0.5\,h^{-1}$ Mpc and $R_2 = 5\,h^{-1}$ Mpc. For this scale $R_s = 1.58\,h^{-1}$ Mpc and $\sigma = 5.78$. Figure 2c shows the same quantities as Figure 2a with the difference that the range of length-scales now corresponds to $R_1 = 1$ and $R_2 = 10\ h^{-1}$ Mpc for which $R_s = 3.16\,h^{-1}$ Mpc and $\sigma = 3.53$. The correspond curves for Millennium Simulations was shown in Figure 2b and 2d respectively.

We may interpret the different curves in Figure 2 as representing the probability of finding a particular kind of structural element in the density environment corresponding to ν. For example, the curve for D=1 in Figure 2a gives the probability of finding a filament in the interval $\nu - 0.25$ to $\nu + 0.25$ for different values of ν. Similarly, the curves for $D = 2$ and $D = 3$ show the probability of finding a sheet and a cluster respectively. Our analysis shows that the filaments, sheets and clusters have different probability distributions. The filaments are preferentially distributed in low density environments relative to the distribution of sheets and clusters. We have a cross-over from this behavior at intermediate densities. The sheets are preferentially distributed relative to filaments and clusters in the high density environments. These findings indicate that the way in which different structural elements are distributed along the Cosmic Web depends jointly on two factors (a) the local density environment, and (b) the length-scale at which we view the Cosmic Web.

In summary the analysis of both the SDSS DR6 and the Millennium Simulation exhibit similar trends. The Local Dimensions were estimated separately in three different ranges of length-scales, 0.5 to 5 h^{-1}Mpc, 1 to 10 h^{-1}Mpc and 5 to 50 h^{-1}Mpc. We find that there is a progressive shift in the D values as we move to larger length-scales. At small length-scales there is a mixtures of sheets and filaments, and the fraction of sheets increases as we move to larger length-scales (1 to 10 h^{-1}Mpc). We find that sheets and clusters are the predominant structures at the largest length -scales (5 to 50 h^{-1}Mpc). Filaments are absent at thesis length-scales. It is interesting to note that [11] find that the mass fraction of filaments decreases while that of sheet increases with an increase in the smoothing length-scale, which is consistent with the results of this paper.

References

[1] Colless M, Dalton G, Maddox S *et al.* 2001 *Monthly Notices of Royal Astronomical Society* **328** 1039–1063 (*Preprint* `arXiv:astro-ph/0106498`)
[2] York D G, Adelman J, Anderson Jr J E *et al.* 2000 *Astronomical Journal* **120** 1579–1587 (*Preprint* `arXiv:astro-ph/0006396`)
[3] Martínez V J and Saar E 2002 *Statistics of the Galaxy Distribution* (Chapman & Hall/CRC press)
[4] Sarkar P and Bharadwaj S 2009 *Monthly Notices of Royal Astronomical Society* **394** L66–L69 (*Preprint* `0812.1800`)
[5] Sarkar P, Pandey B and Bharadwaj S 2011 *ArXiv e-prints* (*Preprint* `1105.4469`)
[6] Springel V, White S D M, Jenkins A *et al.* 2005 *Nature* **435** 629–636 (*Preprint* `arXiv:astro-ph/0504097`)
[7] Sarkar P, Yadav J, Pandey B and Bharadwaj S 2009 *Monthly Notices of Royal Astronomical Society* **399** L128–L131 (*Preprint* `0906.3431`)
[8] Adelman-McCarthy J K, Agüeros M A, Allam S S *et al.* 2008 *Astrophysical Journal Supplement Series* **175** 297–313 (*Preprint* `0707.3413`)
[9] Strauss M A, Weinberg D H, Lupton R H *et al.* 2002 *Astronomical Journal* **124** 1810–1824 (*Preprint* `arXiv:astro-ph/0206225`)
[10] Croton D J, Springel V, White S D M *et al.* 2006 *Monthly Notices of Royal Astronomical Society* **365** 11–28 (*Preprint* `arXiv:astro-ph/0508046`)
[11] Forero-Romero J E, Hoffman Y, Gottlöber S, Klypin A and Yepes G 2009 *Monthly Notices of Royal Astronomical Society* **396** 1815–1824 (*Preprint* `0809.4135`)

Horizon entropy and higher curvature equations of state

Raf Guedens , Ted Jacobson [a], Sudipta Sarkar[a,b]

[a]Center for Fundamental Physics, University of Maryland, College Park, MD 20742-4111, USA. [b]Institute of Mathematical Sciences, Chennai, India

E-mail: `rafguedens@gmail.com, jaconson@umd.edu, sudiptas@imsc.res.in`

Abstract. The Clausius relation between entropy change and heat flux has previously been used to derive Einstein's field equations as an equation of state. We examine here whether a similar derivation can be given for extensions beyond Einstein gravity to include higher derivative and higher curvature terms.

The origin of the thermodynamic behavior of spacetime horizons can be traced to local physics, and in a sense arises from the nature of the vacuum. This led to the observation that the Einstein equation can be derived as an equation of state for local causal horizons in the neighborhood of a point 'p' in spacetime, by imposing the Clausius relation between their entropy change and the energy flux across them [1]. In this paper we present an approach [2], which generalizes this equation of state derivation to allow for higher derivative contributions to the entropy and field equations. The Einstein-Hilbert Lagrangian is only the lowest order term (other than the cosmological constant) in a derivative expansion of generally covariant actions for a metric theory, and the presence of higher derivative terms is presumably inevitable. Hence, it is interesting to examine whether the thermodynamic derivation of the field equations transcends beyond general relativity (GR).

Our approach differs in several ways from the original derivation for GR, among which are: (i) the entropy is compared on two horizon slices that share a common boundary, (ii) the bifurcation surface lies to the past of the terminal point p and, (iii) the entropy depends on the approximate Killing vector. In particular, it has the same dependence on the approximate Killing vector as have the Noether charges associated with a Lagrangian, in analogy with the Wald entropy [3, 4] for stationary black holes. It makes some sense that the entropy depends on the approximate Killing vector, because the latter determines the notion of stationarity and enters the definition of the heat flux. However, at present we can offer no statistical interpretation for this form of the entropy.

The equation of state we can derive is consistent with local energy-momentum conservation only if the leading order term in the entropy satisfies an integrability condition. This condition is satisfied if the entropy arises from variation of a generally covariant function with respect to curvature. In other words, the entropy coincides with a Noether charge associated with a (particular type of) gravitational Lagrangian. The need for such an integrability condition was anticipated in Ref. [1], since it was known that the entropy of stationary black hole horizons has this form[4, 5, 6]. We have not been able to ascertain whether this is the only way to satisfy the integrability condition, and in particular whether field equations for Lagrangians involving derivatives of curvature can be obtained.

To present the derivation, let us first define the notion of a local causal horizon H as follows. Consider any spacetime point p in a D-dimensional spacetime, and let Σ_p be any small patch of spacelike $(D-2)$-surface through p. The boundary of the past of Σ_p in the neighborhood of p has two components, each of which is a null surface generated by a congruence of null geodesics orthogonal to Σ_p. The local causal horizon (LCH), denoted by H is defined as one of these components.

Next we define the approximate Killing vector ξ^a that plays a central role. We will refer to this vector field as the "local Killing vector". Of course a general curved spacetime has no Killing vectors. Nevertheless, in a small enough neighborhood of any point, any spacetime is approximately flat. Then, there exits a notion of an approximate Killing vector ξ^a, such that it obeys properties of a true Killing vector in an approximate way. A true Killing vector ξ^a obeys the Killing equation, $\nabla_{(a}\xi_{b)} = 0$ and also the Killing identity,

$$\nabla_a \nabla_b \xi_c = R^d{}_{abc}\xi_d, \tag{1}$$

which follows from the Killing equation. Our computations will rely on this identity being satisfied at a certain order, but for the approximate Killing field, the Killing identity will generally not hold at that order. Then, narrowing our sights, ξ^b can be chosen so that the Killing identity holds to the required order on the single horizon generator Γ that ends on the terminal point p, and this turns out to be good enough. In effect, the local Killing symmetry can be extended away from p to a better approximation along a single null generator Γ than across the whole LCH.

Given that, we confine the local causal horizon to a narrow region surrounding the central horizon generator Γ, the integrals appearing in the Clausius relation will be dominated by their integrands evaluated on Γ. As such, any condition that ξ^a may need to satisfy in order for the Clausius relation to lead to a consistent equation of state will be conditions imposed on Γ. The local Killing vector ξ^a is taken to have a bifurcation surface Σ_0—or at least a bifurcation point—at p_0 to the past of p, where it vanishes and where its covariant derivative generates boosts in the plane orthogonal to the bifurcation surface. Then, we use the null normal coordinate (NNC) system introduced in [7] to express the full set of requirements that must be imposed on ξ^a and we can show by explicit construction in NNC that both Killing equation and Killing identity are satisfied by the approximate Killing vector ξ^a at the required order as long as we concentrate on a narrow patch near the central null generator Γ of LCH. For details, see Ref. [2].

We also need to define the heat flux through the LCH as,

$$\delta Q = \int_H (-T_{ab})\xi^b dH_a \tag{2}$$

The integration measure is $dH_b = -k_b dV dA$, where V is the affine parameter along the generators of LCH, $k^b = (\partial_V)^b$ is tangent to the generators, and dA is the area element of a constant V slice. The heat flux integrand is proportional to the local Killing vector which vanishes at p_0 and is thus of $O(x)$ in NNCs in the neighborhood of p_0. Also, the relevant temperature is the Unruh boost temperature, $T = \hbar/2\pi$.

The Clausius relation $T\delta S = \delta Q$, applied to a horizon refers to the entropy change δS between two times. Those times are spacelike hypersurfaces, one to the future of the other, which intersect the horizon in two slices. For a black hole with a compact horizon, the two slices may bound a cylindrical region of the horizon. For a local causal horizon, the considered process must be local, since the LCH is not even well defined except in a small neighborhood of the terminal point p. To localize the process we can restrict attention to cases where, rather than pushing the spacelike hypersurface forward in time everywhere, it is deformed to the future only in a small neighborhood of the a bifurcation point p_0. Then the two corresponding horizon slices also coincide everywhere except in a small region. Σ_0 corresponds to the $V = V_0$ surface in NNC of [7] and Σ lies to the future. If we truncate the horizon slices outside the region where they differ, their union $\Sigma_0 \cup \Sigma$ forms the closed boundary of a patch of the horizon. This will allow the difference of the entropies on the two slices to be computed using Stokes' theorem.

In the context of horizon thermodynamics it is natural to expect that the entropy is an extensive quantity. Thus we assume that it can be expressed as an integral of a $(D-2)$-form over a spacelike slice Σ of the horizon. We adopt the dual description, and express the entropy as the integral

$$S = \int_\Sigma s^{ab} N_{ab} dA, \tag{3}$$

where the entropy density s^{ab} is an antisymmetric tensor, N^{ab} is the binormal to the slice and dA is the area element on the slice. The change of this entropy between the two horizon slices is

$$\delta S = S - S_0 = \oint_{\Sigma \cup \Sigma_0} s^{ab} N_{ab} dA = -2 \int_H \nabla_b s^{ab}\, k_a dV dA, \tag{4}$$

where Σ_0 is taken with the opposite orientation to Σ, and in the last step, we have used Stokes' theorem.

The Clausius relation is imposed in the limit $p_0 \to p$, which means that the entropy change integrand of (4), multiplied by T, must be equal to the $O(x)$ heat flux integrand, up to $O(x^2)$ terms. That is,

$$-(\hbar/\pi)\nabla_b s^{ab}\, k_a = T^{ab} \xi_b\, k_a + O(x^2). \tag{5}$$

This relation is imposed at all points p and for all null directions at p.

Next, we propose a local entropy density of the same form as the Noether potential [3, 4] and assume the entropy density takes the form

$$s^{ab} = \frac{2\pi}{\hbar}\left(W^{abc}\xi_c + X^{abcd}\nabla_c \xi_d\right), \tag{6}$$

where W and X are, so far, unspecified tensors, constructed locally from the dynamical fields, antisymmetric in the first two indices, and in addition X is antisymmetric in the last two indices. Then, the validity of Clausius relation $T\delta S = \delta Q$ requires that W is a combination of divergences of X as,

$$W^{arb} = \nabla_s \left(X^{sarb} + X^{sbra} + X^{srba}\right). \tag{7}$$

For details, see Ref. [2]. Using this result and imposing the validity of the Clausius relation (5) for all k^a ultimately imply,

$$R^{(a}{}_{rst} X^{b)rst} + 2\nabla_r \nabla_s X^{(a|s|b)r} + \Phi g^{ab} = -\frac{1}{2} T^{ab}, \tag{8}$$

where Φ is some scalar function that may depend on the metric and curvature. Note that (8) follows from the Clausius relation irrespective of the values of expansion and shear anywhere on the horizon patch.

We now impose local conservation of energy-momentum tensor, T_{ab} to determine the function Φ. Then Eqn. (8) leads to

$$\nabla^a \Phi = -\nabla_b \left(R^{(a}{}_{rst} X^{b)rst} + 2\nabla_r \nabla_s X^{(a|s|b)r}\right). \tag{9}$$

In order for such a Φ to exist, the right hand side must be the gradient of a scalar. This integrability condition further constrains the nature of X, which so far is only required to be anti-symmetric in both the first and second pair of indices.

If the integrability condition is satisfied, then the left hand side of (8) is a divergence free tensor constructed from the metric and its derivatives. One way to obtain such a tensor is from the variational derivative of a scalar action functional with respect to the metric, $\delta I_g[g]/\delta g_{ab}$, which is automatically divergence free. In fact, it was argued in Ref. [8] that *all* such tensors arise in this way. If so, then the 'Clausius equation' (8) is precisely the equation of motion that derives from the action $I_g + I_{\mathrm{matter}}$, with an undetermined cosmological constant. More specifically, if a gravitational Lagrangian $L[g_{ab}, R^a{}_{bcd}]$ is a scalar formed algebraically from the metric and Riemann tensor, then the corresponding equation of motion is precisely (8), with

$$X^{abcd} = -\frac{\partial L}{\partial R_{abcd}}, \tag{10}$$

and $\Phi = L/2$. This completes the derivation of the field equations from the Clausius relationsip applied to LCH.

To summarize, we assume that the entropy of LCH depends on an approximate local Killing vector in a way that mimics the diffeomorphism Noether charge that yields the entropy of a stationary black hole. We showed that by a careful choice of the nature of the horizon patch to which the Clausius relation is applied, it is possible to derive the field equation of a gravity theory whose lagrangian is a algebraic function of metric and curvature. In particular, the Clausius relation must refer to the change of entropy between two slices of the horizon that together form the complete boundary of a patch, and this patch must be narrow enough to neglect violations of the Killing identity.

Together with matter energy conservation, the Clausius relation applied to all such local horizon patches leads to an integrability condition on the assumed horizon entropy density. This condition can be satisfied if the latter is in fact a Noether potential associated with a Lagrangian constructed algebraically from the metric and Riemann tensor. In that case the Clausius relation implies that the field equation for that Lagrangian holds. We have not proved that this is the only way to satisfy the integrability condition, but that may be the case. In particular, the field equation for a theory with derivatives of curvature in the Lagrangian is unlikely to be obtained in this way using for the entropy density a Noether potential derived from the Lagrangian, although it might conceivably arise from a different entropy density.

It is therefore not clear to us whether the derivation has any thermodynamic significance. We regard it as a positive answer to a technical question, but its physical interpretation remains obscure. Then again, it may just be that the contribution of higher curvature terms to a gravitational field equation cannot be sensibly captured at a local thermodynamic level.

References

[1] T. Jacobson, "Thermodynamics of space-time: The Einstein equation of state," Phys. Rev. Lett. **75**, 1260 (1995) [arXiv:gr-qc/9504004].
[2] R. Guedens, T. Jacobson and S. Sarkar, Phys. Rev. D **85**, 064017 (2012) [arXiv:1112.6215 [gr-qc]].
[3] R. M. Wald, "Black hole entropy is the Noether charge," Phys. Rev. **D48**, 3427-3431 (1993). [gr-qc/9307038].
[4] V. Iyer and R. M. Wald, "Some properties of Noether charge and a proposal for dynamical black hole entropy," Phys. Rev. D **50**, 846 (1994) [arXiv:gr-qc/9403028].
[5] M. Visser, "Dirty black holes: Entropy as a surface term," Phys. Rev. **D48**, 5697-5705 (1993). [hep-th/9307194].
[6] T. Jacobson, G. Kang, R. C. Myers, "On black hole entropy," Phys. Rev. **D49**, 6587-6598 (1994). [gr-qc/9312023].
[7] R. Guedens, arXiv:1201.0542 [gr-qc].
[8] E. Curiel, "On Tensorial Concomitants and the Non-Existence of a Gravitational Stress-Energy Tensor," arXiv:0908.3322 [gr-qc].

AUTHOR INDEX